CHAOS ANALYSIS AND CHAOTIC EMI SUPPRESSION OF DC-DC CONVERTERS

CHAOS ANALYSIS AND CHAOTIC EMI SUPPRESSION OF DC-DC CONVERTERS

Bo Zhang and Xuemei Wang

South China University of Technology, P. R. China

IEEE

WILEY

This edition first published 2015

Registered office
John Wiley & Sons Singapore Pte. Ltd., 1 Fusionopolis Walk, #07-01 Solaris South Tower, Singapore 138628.

For details of our global editorial offices, for customer services and for information about how to apply for permission to reuse the copyright material in this book please see our website at www.wiley.com.

Library of Congress Cataloging-in-Publication Data

Zhang, Bo, 1962-
Chaos analysis and chaotic EMI suppression of DC-DC converters / Bo Zhang, Xuemei Wang.
pages cm
Includes bibliographical references and index.
ISBN 978-1-118-45100-7 (hardback)
1. DC-to-DC converters. 2. Electromagnetic interference–Prevention. 3. Chaotic behavior in systems. I. Wang, Xuemei (Electrical engineer) II. Title.
TK7872.C8Z433 2014
621.31′32 – dc23

2014022263

Typeset in 11/13pt TimesLTStd by Laserwords Private Limited, Chennai, India

1 2015

M 03M000030000005241 120422

Contents

About the Authors

Dr Bo Zhang was born in Shanghai, China, in 1962. He received a B.S. degree in electrical engineering from Zhejiang University, Hangzhou, China, in 1982; an M.S. degree in power electronics from Southwest Jiaotong University, Chengdu, China, in 1988; and a Ph.D degree in power electronics from Nanjing University of Aeronautics and Astronautics, Nanjing, China, in 1994.

He is currently a Professor and the Vice Dean with the School of Electric Power, South China University of Technology, Guangzhou, China. He has authored or coauthored more than 350 papers and 17 patents. His current research interests include nonlinear analysis and control of power electronics and AC drives.

Dr Xuemei Wang was born in Hunan, China, in 1972. She received her Ph.D degree from the South China University of Technology, Guangzhou, China, in 2009.

She is currently an Associated Professor with the School of Electric Power, South China University of Technology, Guangzhou. Her main research interests include nonlinear analysis and control of power electronics.

Preface

This book focuses on the analysis and application of chaos to reduce harmful EMI (electromagnetic interference) of DC–DC converters. DC–DC converters are typical switching systems which have plenty of nonlinear behaviors, such as bifurcation and chaos. The nonlinear behaviors of DC–DC converters have been studied in depth over the last 30 years; in contrast, people are still puzzled by the practical applications of bifurcations and chaos in switching converters. The EMI, due to the high rates of changes of voltage and current resulting from the switching process of power semiconductors, has been a major design constraint of DC–DC converters for a long time. How to reduce the annoying, harmful EMI has consistently attracted much research interest. The conventional filtering and shielding approaches have disadvantages mainly in cost and weight. Is there any method of suppressing EMI which is simple, ingenious, and may solve the EMI problem fundamentally?

Pulse width modulation (PWM) control is the most common and important control method for switching converters. Over the last 10 years, researchers have concerned themselves with applying frequency modulation technology to the suppression EMI of switching converters. The basic principle of frequency modulation makes the EMI energy distribution uniform and reduces the peak values of spectrum, thus suppressing EMI. Some commonly used frequency modulation techniques are jitter frequency, periodic frequency modulation, random frequency modulation, and so on. Comparing these frequency modulation techniques, it is recognized that chaotic frequency modulation may reduce the harmonic greatly and improve the spectral distribution because the chaos signal has a prominent characteristic of continuous power spectral density. Combining the chaos and PWM control technique can distribute the harmonics of the DC–DC converters continuously and evenly over a wide frequency range, consequently the EMI may be reduced.

Although chaotic frequency modulation has significant advantages, this technology has not been applied in industrial products because there are two barriers between the theory research and applications. One is that the current research results of chaos are almost all abstract mathematical presentations baffling to power electronics engineers. Another is that the spectral feature of the chaotic signal is inner-harmonics and the non-integral multiplies harmonics which are difficult to estimate and quantify in the traditional Fast Fourier Transform (FFT) method. Toward the first objective, analysis approaches associated with chaos phenomena in power electronics, which are easy to understand and employ, are needed. For the second objective, it is necessary to adapt some available method beyond FFT to express the chaos power spectral density correctly. So, there are two parts in this book. The first part, including Chapters 2–4, is concerned with a series of the new chaos quantified method of DC–DC converters; the second part, including Chapters 5–10, is about chaotic power spectral density estimating and chaotic modulating technologies.

A brief outline of the 10 core chapters is as follows. After a review of the fundamentals of chaos behaviors of DC–DC converters, the authors present some recent findings such as symbolic entropy, complexity, and invariant probability distribution, to analyze the characteristics of chaotic DC–DC converters in the next three chapters. Using these methods, the statistic characteristics of chaotic DC–DC converters are extracted and the foundations for the following researches of chaotic EMI suppression are reinforced. The focus then is transferred to estimating the power spectral density of chaotic PWM converters after the introduction of EMI and Electromagnetic Compatibility (EMC) of switching converters. After an introduction to the basic principles of spectrum analysis, chaos point process, Prony, and wavelet analysis methods are suggested for estimating the power spectral density of chaotic PWM converters. Finally, some design-oriented applications which provide a good example of applying chaos theory in engineering practice illustrate the effectiveness in suppressing EMI of the proposed chaotic PWM.

The book highlights the advanced research works in the fields of statistic characteristics of nonlinear behavior and chaotic PWM modulation technology to suppress the EMI of switching converters. The proposed analysis method will directly contribute to the studies of EMI suppression and be available to practical design problems. We think, in the near future, the method and technology proposed in this book will benefit power electronics engineers and will be widely used in power electronics.

Bo Zhang and Xuemei Wang
South China University of Technology
Guangzhou in China
December 2013

Acknowledgments

There are many people to whom I am very obliged for their support.

I am deeply grateful to my former students Dr Ru Yang, Dr Hong Li, Dr Xuemei Wang, and PhD student Fan Xie, for their wonderful work in preparing their PhD theses.

I would like to thank Professor Dongyuan Qiu and PhD student Fan Xie, Dongdong Wang for improving the language and for the creation of some of the plots in this book. I am very obliged to the staff at John Wiley & Sons, especially to Ms Clarissa Lim and Mr Mingxin Hou, for taking care of the preparation of this book to press.

I would like to acknowledge the help of the sustentation fund of the Natural Science Foundation of China (Nos 50937001 and 60474066), the Natural Science Foundation of Guangdong Province (No. 05103540), the National High Technology Research and Development Program of China (No. 2007AA05Z299), and the National Key Technology Support Program (No. 2008BAF34B09). There is no doubt that this book would not have come out without the support from these organizations.

Bo Zhang
South China University of Technology
Guangzhou in China
December 2013

1

Nonlinear Models and Behaviors of DC–DC Converters

1.1 Introduction

DC–DC converters are widely used in industrial and commercial applications with the requirement of regulating DC power, such as computers, spacecrafts, medical instruments, communication devices, electrical vehicles, and so on [1–6]. The technology of DC–DC conversion is a major subject area in the field of power engineering and drives, and has been under development for seven decades [5]. It is well known that DC–DC converters are typical nonlinear systems because of their switching processes. Some irregular behavior, such as subharmonics and intermittent instability, has been observed constantly in practice. Over nearly three decades, lots of nonlinear behavior, such as period-doubling bifurcation, Hopf bifurcation, border collision bifurcation, torus bifurcation, coexistence attractor, intermittent chaos and chaos, has been studied by power electronics researchers one after the other [7–17].

Although the nonlinear behavior of DC–DC converters has been intensively studied, people are still puzzled by the practical application of bifurcations and chaos in switching converters [18]. Miniaturization is the main approach to reduce the size of DC–DC converters, and a higher switching frequency must be utilized with the requirement of the smaller size. Electromagnetic interference (EMI), resulting from bursting changes of voltage or current (dv/dt or di/dt), has been a major design constraint for a long time due to the high switching frequency. The question of how to reduce annoying and harmful EMI has attracted much research interest.

Fortunately, chaos research opens up the possibility of the application of nonlinear characteristics in power electronic systems, and might be a new method to improve the performance of power electronic systems in the future. For example, the switching speed between unstable periodic orbits (UPOs) embedded in chaos attractors is fast. Using this performance, DC–DC converters working in chaos may have a fast dynamic response [19]. Moreover, it has been actually proved that the application of

Chaos Analysis and Chaotic EMI Suppression of DC-DC Converters, First Edition. Bo Zhang and Xuemei Wang.

chaotic power spectral characteristics can effectively reduce the EMI and improve the electromagnetic compatibility (EMC) of DC–DC converters [20–25].

Although spectrum analysis of chaotic switching converters has resulted in some research achievements, there are still fewer quantitative indicators and comprehensive assessments of the characteristics of the chaotic spectrum because the switching converter is a complex nonlinear, nonautonomous, and time varying system. First of all, we need to know about the nonlinear behavior of switching converters and some basic dynamical concepts. In this chapter, we will begin with an introduction to DC–DC converters and their operation modes. Next, the conventional modeling methods are presented. Some basic nonlinear dynamics knowledge is then introduced in the rest of this chapter.

1.2 Overview of PWM DC–DC Converters

1.2.1 *Principle of Pulse Width Modulation*

DC–DC converters supply a regulated DC output voltage to a variable-load resistance effectively. Switch-mode DC–DC converters convert one DC input voltage V_{in} to a desired output voltage V_{out} by storing the input energy temporarily and then releasing that energy to the output load. The average value of voltage is moderated by controlling the switch on and off durations (t_{on} and t_{off}). In Figure 1.1, the input voltage V_{in} is chopped to an output voltage V_{out}. The longer the switch is on compared to the off duration, the higher the power supplied to the load. So, the duty cycle, describing the ratio of on duration to the switching time period T, is defined as

$$d = t_{on}/T \tag{1.1}$$

Obviously, the average output voltage varies depending on d. By varying d, V_{out} can be controlled. So, this method is called *Pulse Width Modulation* (PWM).

The simplest way to generate a PWM signal is by comparison, which requires only a sawtooth or a triangle waveform and a comparator. When the value of the control signal (v_{con}, the dash line, commonly comes from the error of the actual output voltage and reference voltage) is greater than the modulation waveform (v_{ramp}, the sawtooth with fixed frequency T and peak value V_U), the comparator output becomes high,

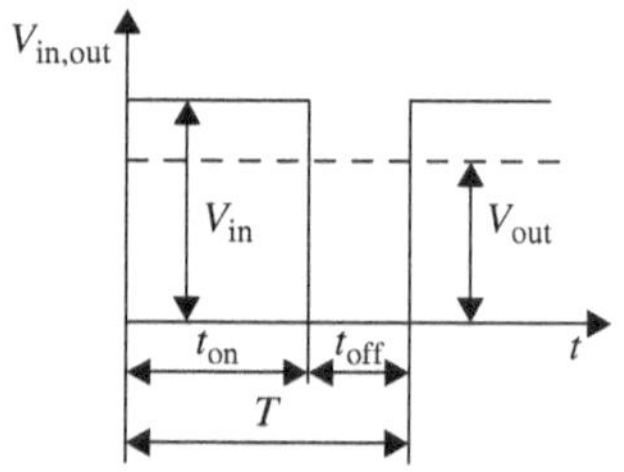

Figure 1.1 DC–DC converter voltage waveforms

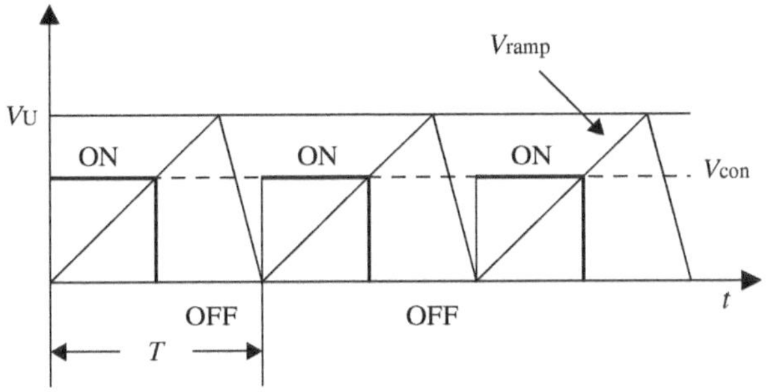

Figure 1.2 Pulse width modulation concept

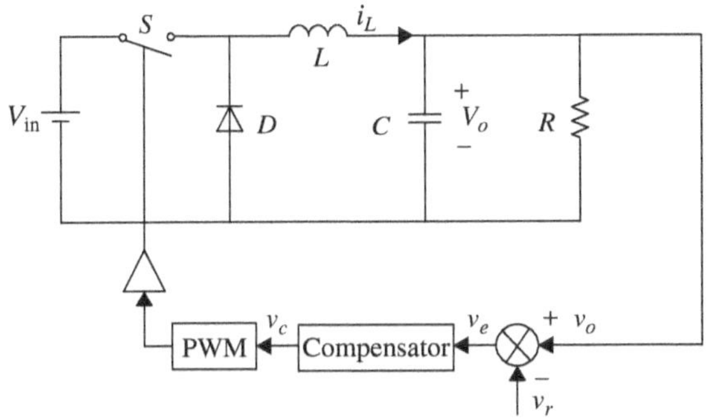

Figure 1.3 Closed-loop controlled PWM system to regulate the output voltage

otherwise it is in the low state shown in Figure 1.2 [4]. In practical terms, a PWM controller consists of three main components: a clock for setting the switching frequency, an output voltage error amplifier, and a sawtooth signal synchronized with the clock.

To illustrate, a PWM step-down converter and feedback loop block is shown in Figure 1.3 as a example. A closed-loop converter contains three ports: a switching main circuit, sampling circuit, and a control circuit that is introduced to regulate the output voltage. It is desired to design the feedback system in such a way that the output voltage is accurately regulated, and is insensitive to disturbances in v_g or in the load current [3]. A control system can be constructed by causing the output voltage to follow a given reference voltage by varying the duty cycle, because the output voltage is a function of the switch's duty cycle.

1.2.2 Basic Topologies of DC–DC Converters

According to incomplete statistics, there have been more than 500 prototypes of DC–DC converters developed over the past seven decades. But all of them come

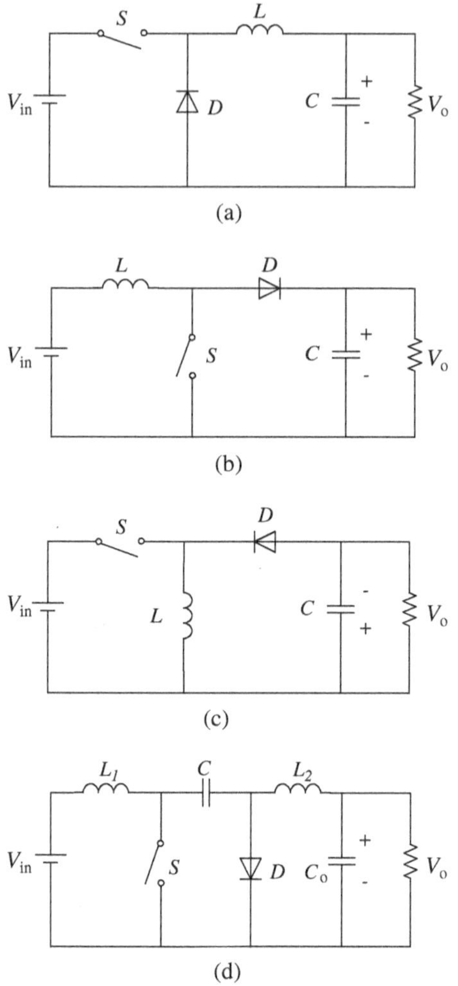

Figure 1.4 Basic topologies of nonisolated DC–DC converters including (a) buck, (b) boost, (c) buck–boost, and (d) Ćuk converters

from the several basic topologies shown in Figure 1.4, where switch *S* and diode *D* are alternately on and off. In these topologies, converters, are the two most basic topologies. The buck–boost converter carries out both the stepping up and down action. The Ćuk converter is a duality buck–boost converter.

If the output is required to be electrically isolated from the input, isolated DC–DC converters, whose isolation is provided by a high frequency isolation transform, are needed. There are two kinds of isolated topologies, unidirectional and bidirectional core excitation based on the way they use the transformer core. The unidirectional

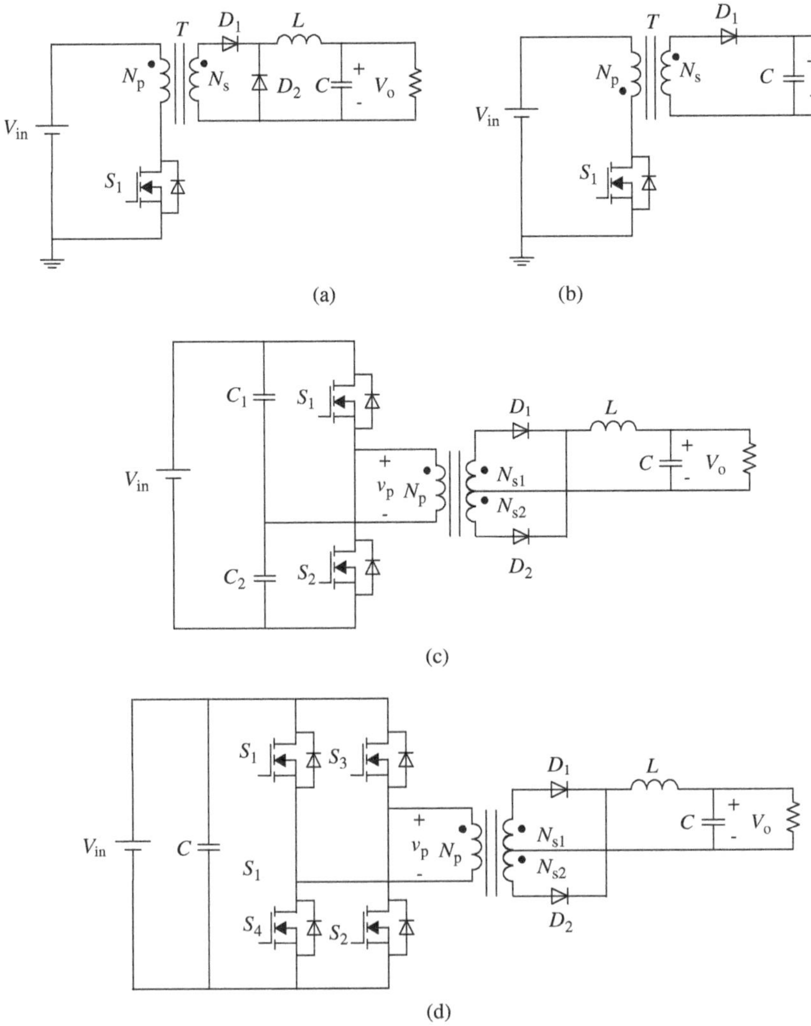

Figure 1.5 Isolated DC–DC converters including (a) forward, (b) flyback, (c) half-bridge, and (d) full-bridge converters

topology has two categories: forward converter and flyback converter, whose output voltages are regulated by means of the PWM scheme. The bidirectional topology has three categories: push-pull, half-bridge, and full-bridge. The circuits of forward, flyback, half-bridge and full-bridge converters are shown in Figure 1.5(a)-(d) respectively. Topologically, the flyback converter is an isolated buck–boost converter, and the half-bridge and full-bridge are isolated buck converters.

1.2.3 *Operation Modes of DC–DC Converters*

DC–DC converters operate in one of two modes depending on the characteristics of the inductor current [2, 4]:

1. *continuous conduction mode* (CCM),
2. *discontinuous conduction mode* (DCM).

As shown in Figure 1.6, the continuous conduction mode is defined by the continuous output current (greater than zero) over the entire switching period, whereas the DCM is defined by the discontinuous output current (equal to zero) during any portion of the switching period. Each mode is discussed in relation to the buck–boost converters in subsequent sections [4].

1.2.3.1 Continuous Conduction Mode

The operation of CCM in steady state consists of two states – switch-on and switch-off modes – whose equivalent circuits of boost converter are illustrated in Figure 1.7. When the switch is on for a time duration t_{on}, the switch conducts the source power and the diode becomes reverse biased, then the inductor current linearly increases. When the switch is off for a time duration t_{off}, the diode becomes forward biased, and the inductor current increases.

1.2.3.2 Discontinuous Conduction Mode

The operation of DCM in steady state consists of three states besides the two modes mentioned above, and the third one is that *S* and *D* are all off and the inductor current

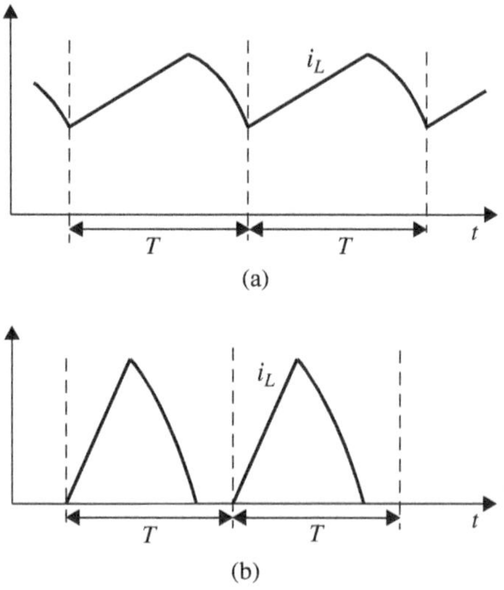

Figure 1.6 Inductor current waveforms at (a) CCM and (b) DCM

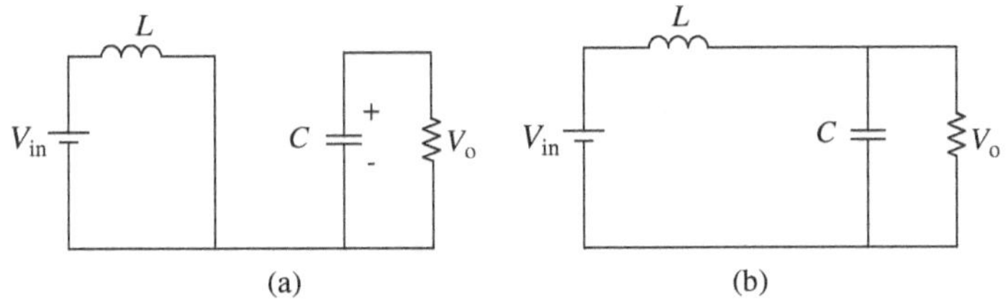

Figure 1.7 CCM circuit states of boost converter when (a) *S* on, *D* off and (b) *S* off, *D* on

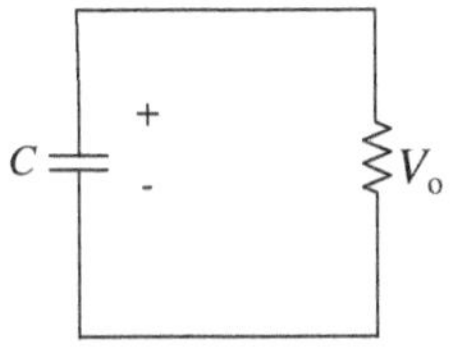

Figure 1.8 DCM circuit state of boost converter (*S* off, *D* off)

stays zero. Figure 1.8 shows the waveforms for the DCM. In this mode, the inductor current drops all the way to zero some time after the switch is turned off, and then remains at zero, with the transistor and diode both off, until the transistor is turned on again.

1.2.4 State-Space Model of DC–DC Converters

A feedback controlled buck converter is illustrated in Figure 1.3. In order to design a stable feedback system, and properties such as transient overshoot, settling time, and steady state regulation should meet special demands, and the model of converter should be established first. The dynamic model of a DC–DC converter can be one of several [2, 4]:

1. circuit model,
2. linearized model,
3. state-space model and averaged state-space model,
4. discrete time sampled data model.

The state-space model is the most popular among those mentioned above, and is the key step in modeling PWM converters with the small-signal linearization model. We will use the buck converter to illustrate the state-space modeling method of DC–DC converters.

A *state-space model* is a model of a system that may be represented by a differential equation. Here the state is a collection of variables summarizing the past of a system for the purpose of predicting the future. The system is called time-invariant if the different equations do not explicitly depend on time t. However, it is possible to have more general time-varying systems because the functions depend on time.

A system is called linear if the functions are linear in $\boldsymbol{x}$ and u. A linear state-space system can thus be represented as state equations in the matrix forms of

$$\begin{cases} \dot{\mathbf{x}} = A\mathbf{x} + Bu \\ y = C\mathbf{x} + Du \end{cases} \tag{1.2}$$

where $\mathbf{x}$ is the *state variables*, which are a minimum number of variables to uniquely specify the state of the system, u is the independent input to the system, A, B, C, and D are constant matrices. Such a system is said to be *linear time invariant* (LTI).

For the buck DC–DC converter, choosing the inductor current and capacitor voltage as natural state variables, and picking the input voltage source as u and resistor voltage as the output v_o, there are two state matrices for CCM operation based on the circuit theory.

Mode 1: Switch S is on, at duration dT:

$$\begin{cases} \dot{\mathbf{x}} = A_1\mathbf{x} + B_1u \\ v_o = C_1\mathbf{x} \end{cases} \tag{1.3}$$

Mode 2: Switch S is off, at duration $(1-d)T$:

$$\begin{cases} \dot{\mathbf{x}} = A_2\mathbf{x} + B_2u \\ v_o = C_2\mathbf{x} \end{cases} \tag{1.4}$$

Here, $\mathbf{x} = [v_o \;\; i_L]^T$, $A_1 = A_2 = \begin{pmatrix} 0 & -\frac{1}{L} \\ \frac{1}{C} & -\frac{1}{RC} \end{pmatrix}$, $B_1 = [1/L \;\; 0]^T$, $B_2 = [0 \;\; 0]^T$, and $C_1 = C_2 = [0 \; 1]^T$.

The *state-space averaged model* is a method of deriving the average value of state variables in one period. For switching converters, averaging and weighting the state variables by duty cycle over a switching period T, we obtain a state-space averaged equation

$$\begin{cases} \dot{\mathbf{x}} = \left[A_2d + A_2(1-d)\right]\mathbf{x} + [B_1d + B_2(1-d)]u \\ v_o = [C_1d + C_2(1-d)]\mathbf{x} \end{cases} \tag{1.5}$$

Next, disturbance and linearization methods are utilized near the working points to obtain a small-signal model. Finally, the transfer function may be obtained using the small-signal model to design the linear controller through classic control theory.

Generally speaking, although the proponents of a given method may prefer to express the end result in a specific form, the end results of nearly all methods are equivalent.

1.2.5 *Discrete Model of DC–DC Converters*

Under the assumption of a converter operating in a small range beside the steady-state working points, linear models are used extensively to describe the dynamics of a system. However, when a system is required to operate over a large range, in most cases, linear models do not provide satisfactory descriptions of the system's behavior. The discrete model is a kind of large signal model.

If we observe the representative points of every fixed interval, unlike the continuous motions of the orbits in phase plane, a series of discrete points may be obtained. And studying the motion of the orbits comes down to the motion of discrete points. The time interval is called the sampling period, and the discrete points are called sampling points. Mapping by sampling constant time intervals is called *stroboscopic mapping*.

Example 1.1 A Boost Converter A schematic diagram of a peak-current control boost converter is shown in Figure 1.9. The current waveform is illustrated in Figure 1.10. It is assumed that the boost converter is working in CCM, so that the inductor current never falls to zero. When the current through inductor L reaches the reference peak current I_{ref}, the comparator outputs a high level voltage which may reset the RS-flipflop to a low voltage and drive the switch S off, while the inductor releases energy that makes the inductor current decrease. When the next clock pulse comes, the RS-flipflop will be set to a high voltage and drive the switch S on to restore energy to the inductor, and the inductor current increases.

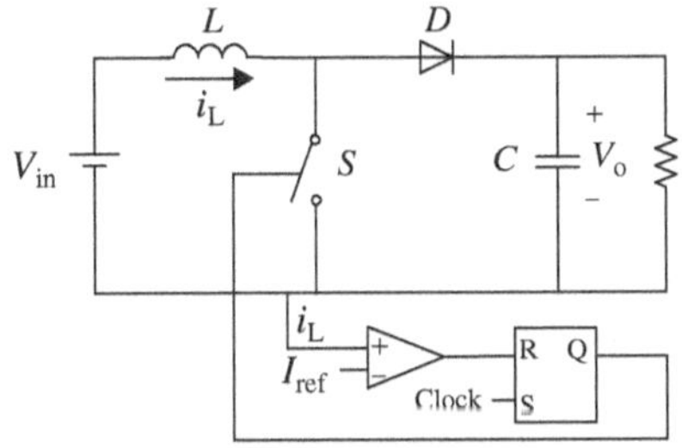

Figure 1.9 Schematic diagram of boost converter of peak-current control

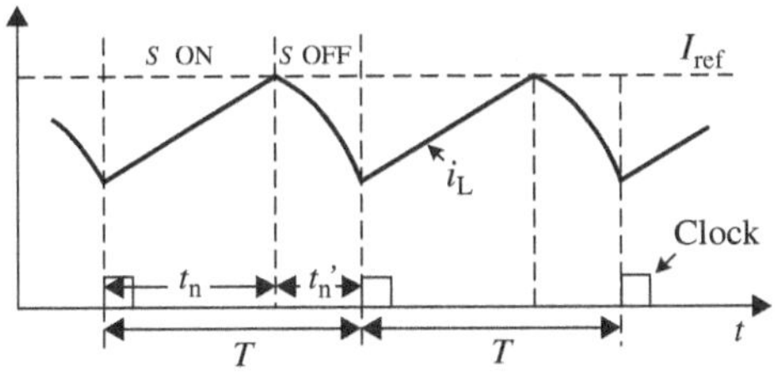

Figure 1.10 Current waveform of peak-current control

Stroboscopic maps relate the time-discrete state vector at every clock instant [19, 20]. Let i_L and v_C at the nth switching moment be i_n and v_n respectively, and those at the next moment be i_{n+1} and v_{n+1} respectively. We get

$$x_{n+1} = \begin{cases} e^{A_2(T-d_nT)}\left(e^{A_1 d_n T}x_n + \int_0^{d_nT} e^{A_1(d_nT-\tau)}d\tau B_1 E\right) & \\ \quad + \int_{d_nT}^{T} e^{A_2(T-d_nT-\tau)}d\tau B_2 E) & d_n < T \\ e^{A_1T}x_n + \int_0^T e^{A_1(T-\tau)}d\tau B_1 E & d_n \geq T \end{cases} \tag{1.6}$$

If A_1 and A_2 are all full rank,

$$x_{n+1} = \begin{cases} e^{A_2(T-d_nT)}\left[e^{A_1 d_n T}x_n + \left(e^{A_1 d_n T} - I\right)A_1^{-1}B_1 E\right] & \\ \quad + \left(e^{A_2(T-d_nT)} - I\right)A_2^{-1}B_2 E & d_n < T \\ e^{A_1T}x_n + (e^{A_1T} - I)A_1^{-1}B_1 E & d_n \geq T \end{cases} \tag{1.7}$$

Using Equation 1.7, the stroboscopic map of a peak-current control boost converter is [27]:

$$\begin{cases} i_{n+1} = e^{-kt'_n}\left(A_1 \sin\omega_1 t'_n + A_2 \cos\omega_1 t'_n\right) + E/R \\ v_{n+1} = E + e^{-kt'_n}\left[(A_1 kL + A_2\omega_1)\cdot \sin\omega_1 t'_n + (A_2 k - A_1\omega_1)L\cos\omega_1 t'_n\right] \end{cases} \tag{1.8}$$

where, $k = 1/2RC$, $\omega_1 = \sqrt{1/LC - R^2}$, $t_n = L(I_{\text{ref}} - i_n)/E$, $t'_n = T - t_n$, $A_1 = \frac{kL(I_{\text{ref}}-E/R)+E-v_n e^{-2kt_n}}{\omega_1 L}$, and $A_2 = (I_{\text{ref}} - E/R)$.

In this example, the sampling time instants of the stroboscopic map are synchronous with the switching events, thus it is also called a *stroboscopic switching map*. Moreover, for voltage-mode control DC–DC converters, there is always a special phenomenon called the *skipped cycle* in the case that the control voltage is beyond the ramp voltage. Fortunately, S-switching and A-switching maps are utilized to model DC–DC converters for this particular situation. Detailed introductions to S-switching and A-switching maps are described in [28], and they are not presented again in our book.

1.3 Overview of the Nonlinear Behavior of DC–DC Converters

In this section, we will take a look at the nonlinear phenomena that govern the complex motion of DC–DC converters. Studies of the nonlinear behavior of power electronic converters began with DC–DC converters over the past 30 years. In 1984, the chaos phenomenon of the buck converter was first mentioned by Brochett and Wood [29]. Hamill and Jefferies [30] introduced bifurcation and chaos in a PWM buck converter, where the difference equations and return maps were utilized to analyze its stability domain in 1988. The phenomena of boundedness, intermittency, and chaos were then observed in an experiment by Krein and Bass in 1990 [31]. From then on, plenty of complex behavior was discovered in power electronic converters.

We give an example of a boost converter in peak current control in Figure 1.9 to illustrate several basic methods and the usual nonlinear behavior of DC–DC converters.

A *phase portrait* is a collection of trajectories that represents the solutions of equations in the state-space. For instance, using the parameters provided gives an example of phase portraits corresponding to time domain waveforms as shown in Figure 1.11 [27].

A qualitative change in the number of solutions to a dynamical system by varying a parameter is called a *bifurcation*. A *bifurcation diagram* is a graphical representation of bifurcation: a parameter is varied and plotted along the *x*-axis, and the asymptotic

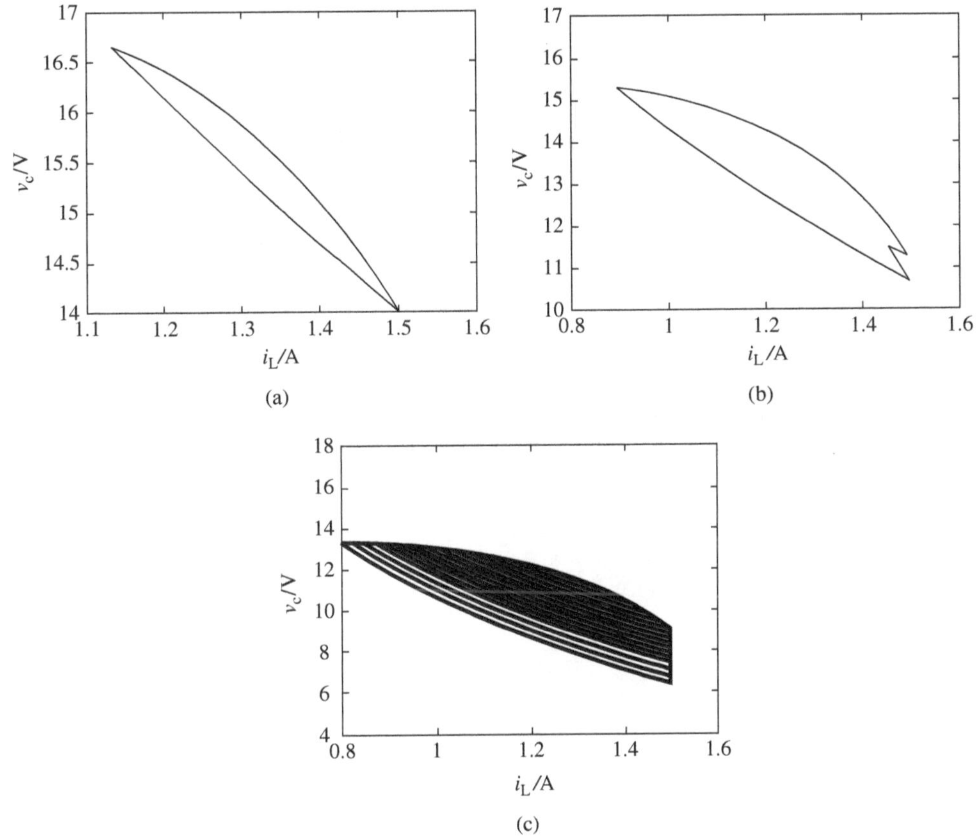

Figure 1.11 Phase portraits of (a) period-1, (b) period-2, and (c) chaos. Reprinted, with permission, from Deane, J.H.B, Chaos in a current-mode controlled dc-dc converter, IEEE Trans. Circuits Syst. I, and Aug./1992

behavior of a sampled state variable is plotted on the y-axis as discrete points [28]. A bifurcation diagram of a boost converter is shown in Figure 1.12. If the system is operating in period-l for some parameter values corresponding to a cycle in the phase portrait, there will be only one point in the bifurcation diagram. If it is in period-2, there will be two points. Such a bifurcation diagram summarizes the change in system behavior in response to the variation of a parameter [28].

Decreasing the value of converter's input voltage E, the system successively undergoes period-1, period-2, period-4, ..., period-2n (n is the positive integers) and chaotic orbits as shown in Figure 1.12. The discrete points filled in in the bifurcation diagram are obtained by sampling the inductor current i every T time interval repeatedly, where T is the switching period.

In this converter, there are two kinds of bifurcations, period-doubling bifurcation and border-collision bifurcation. Period-doubling bifurcation is a smooth and standard bifurcation, whose period number increases multiply as period-1, period-2, period-4, and so on, which is common in buck or boost converters with current-mode controlled and voltage-mode controlled [18].

As we know, for DC–DC converters, the duty cycle of a switch is limited in 0–1. Suppose that the converter operates in CCM; three kinds of possible current waveforms within one clock period T with the current-peak controlled are shown in Figure 1.13. The stable current waveform usually undergoes two working modes, mode 1 and mode 2, introduced previously, Figure 1.13a, where its duty cycle is $0<d<1$. There is another unsteady state presented in Figure 1.13b, where only mode 1 appears, and whose duty cycle d is saturated and equals 1. If the current waveform just touches the reference current at the end of clock T, Figure 1.13c, it is a boundary situation between Figure 1.13a,b. The border-collision bifurcation exists due to the saturation of the duty cycle, and occurs when duty cycle d reaches one of the saturation values 0 or 1. This type of bifurcation occurs in dynamical

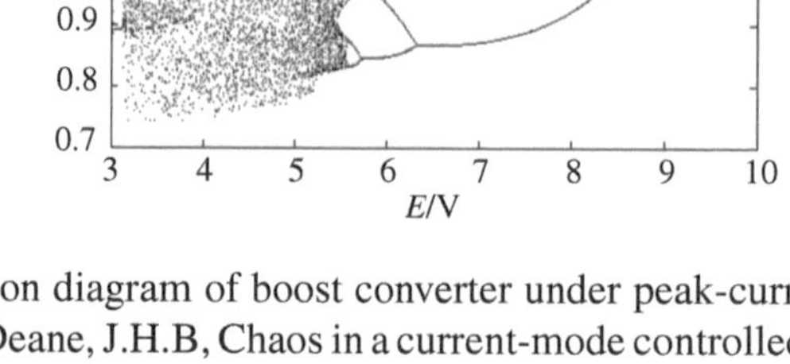

Figure 1.12 Bifurcation diagram of boost converter under peak-current control. Reprinted, with permission, from Deane, J.H.B, Chaos in a current-mode controlled dc-dc converter, IEEE Trans. Circuits Syst. I, and Aug./1992

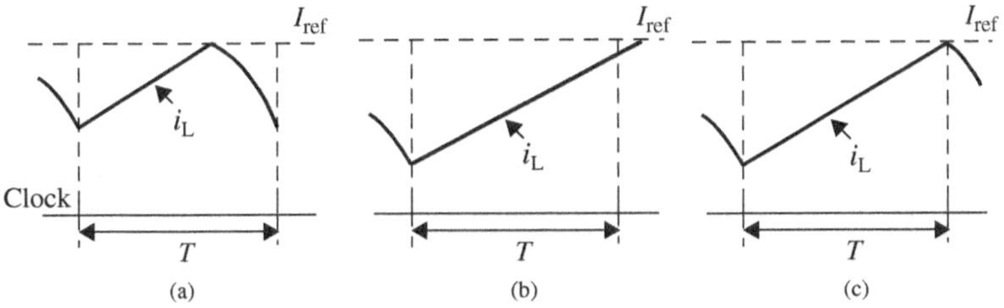

Figure 1.13 Modes of CCM of peak-current control at (a) $0 < d < 1$, (b) $d = 1$, and (c) boundary situation

systems where two or more structurally different systems operate for different parameter ranges [18]. Mathematically, when orbits of a piecewise-smooth system (introduced subsequently) tangent to some borders, this bifurcation will occur. We might distinguish this for the transform from period-2 to period-4, there is an unsmooth curve-like fold at the top of the bifurcation diagram which means the appearance of border-collision bifurcation. Of course, border-collision bifurcation is a nonstandard bifurcation.

In Figure 1.11c, we find the trajectory appears to move randomly in the state-space. Moreover, the trajectory is bounded and the motion is nonperiodic [18], or termed bounded aperiodic orbits. In the bifurcation diagram, we also find that there are plenty of populated points (infinite theoretically), which look like snow on a television set. We may roughly discover some properties as follows: (i) it looks random because the points are dense and chaotic (ii) but, it is ordered because the points concentrate in a certain bound. Roughly speaking, trajectories of the system that start in a particular bounded region of the state-space (known as *chaos* or *strange attractor*) will forever remain in that region, without exhibiting any apparent regularity or pattern (e.g., without converging to an equilibrium or becoming periodic).

1.4 Review of Basic Dynamics Concepts

Since Newton mechanics was founded, people have firmly believed that if a deterministic input is imposed on a deterministic dynamical system, then the output of the system must be determined as well. For the linear system, this conclusion is quite correct, but for the nonlinear system, some motions that may not be precisely repeated and are seemingly random may emerge.

Unfortunately, PWM DC–DC converters are none other than this kind of nonlinear system. The usage of semiconductor switches means that PWM DC–DC converters represent a hybrid system, which involves both continuous-valued and discrete-valued variables. Therefore, due to the nature of the switching process, PWM DC–DC converters are not a continuously differentiable system nor a nonsmooth system, which may give rise to complex dynamics. In this section, we will start with some basic concepts of a dynamical system from dynamics theory. This material is available in many textbooks on nonlinear dynamics [33–37]. So we shall always try to present their concepts in geometrical images for ease of understanding.

1.4.1 Dynamical System

Roughly speaking, a dynamical system describes the evolution of a state over time. The evolution rule of the dynamical system is a fixed rule that describes what future states follow from the current state. According to the form of state evolution, dynamical systems can be classified into:

1. *continuous-time dynamical systems* and
2. *discrete-time dynamical systems.*

1.4.1.1 Continuous-Time Dynamical System

The state of continuous-time dynamical systems is allowed to change (continuously or discontinuously) at all times t. *Ordinary differential equations* (*flows*) are employed to describe continuous-time dynamical systems. A generalized continuous-time dynamical system is governed by the state-space model in vector form

$$\dot{\mathbf{x}} = f(\mathbf{x}, t) \tag{1.9}$$

where $\mathbf{x}$ is the real, nth-order state vector in $\mathbb{R}^n$. $\dot{\mathbf{x}}$ is called a vector field, it gives a vector pointing in the direction of the velocity at every point in phase space.

If there is no time term t at the right hand side of Equation 1.9

$$\dot{\mathbf{x}} = f(\mathbf{x}) \tag{1.10}$$

this is called a form of *autonomous system.* For the actual system, it means that the system does not have any externally applied time-varying input or other time variation. For example, a three-dimensional Lorenz system is

$$\begin{cases} \dot{x}_1 = a(x_1 - x_2) \\ \dot{x}_2 = (1 + b)x_3 - x_2 - x_1x_3 \\ \dot{x}_3 = x_1x_2 - cx_3 \end{cases} \tag{1.11}$$

where a, b, c are parameters.

Conversely, Equation 1.9 is a *nonautonomous system* which has external inputs or forcing functions or time variations. For example, PWM power electronic circuits are explicitly time-dependent nonautonomous systems whose operations are compulsively controlled by clock logic with period T.

For a continuous-time dynamical system, the basic geometrical objects associated with a dynamical system are its orbits in the state-space and the phase portrait composed of these orbits.

Equilibrium: Consider the autonomous system, a state $\mathbf{x}^* \in \mathbb{R}^n$ is called an equilibrium of Equation 1.9 if $f(\mathbf{x}^*) = 0$.

Cycle: A cycle is periodic orbit for a vector field, namely a nonequilibrium orbit.

Limit cycle: A cycle of a continuous-time dynamical system, in a neighborhood of which there are no other cycles, is called a limit cycle [31].

Obviously, a limit cycle is the periodic solution with period T satisfying the equation

$$x(t+T) = x(t)$$

The trajectory in phase-space is a point or a closed curve, and is called an equilibrium or limit cycle as shown in Figures 1.14a,b respectively. Physically, the stable limit cycle is called self-oscillation.

1.4.1.2 Discrete-Time Dynamical Systems

The state of discrete-time dynamical systems can only change at discrete time instances. Difference equations are employed to describe discrete-time dynamical systems. In some circumstances, it is more natural to describe the evolution of a system at discrete instants of time rather than continuously in time, for example, in ecology and economics when the state of a system at a certain moment of time t completely determines its state after a year, say at $t+1$. Discrete time refers to the set of times that is a subset of the integers. We will use $n \in \mathrm{Z}$ to denote discrete time. Typically, the evolution of the state of a discrete time system with a control parameter is expressed by a difference equation [33, 37]

$$\mathbf{x}_{n+1} = f(\mathbf{x}_n, \mu) \tag{1.12}$$

where $\mathbf{x}$ is the nth-dimension state variable, μ is the set of all control variables, and this equation also may be called a map f, written as $x \mapsto f(x)$.

A *fixed point* of the discrete-time dynamical system is $\mathbf{x}^*$ such that $\mathbf{x}^* = f(\mathbf{x}^*, \mu)$. Defining $f^n(\mathbf{x}_0)$ is the result of n times iterations of Equation 1.12 from an initial $\mathbf{x}_0$. If finally $f^n(\mathbf{x}_0) = \mathbf{x}_0$, and for an arbitrary integer $k < n$, $f^k(\mathbf{x}_0) \neq \mathbf{x}_0$ is always satisfied,

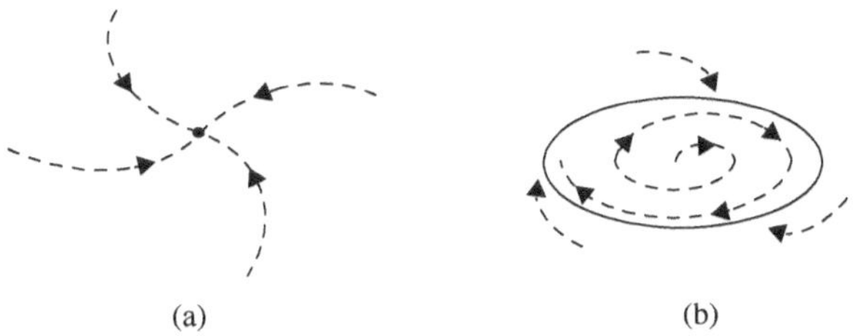

Figure 1.14 Illustration of stable (a) equilibrium and (b) limit cycle in a continuous-time system

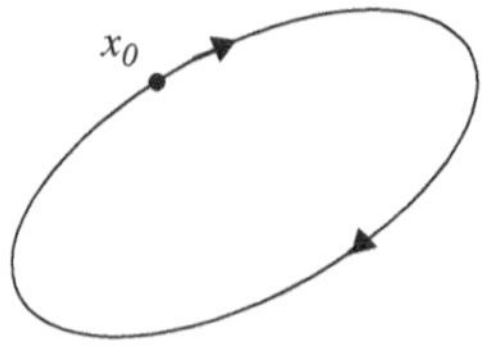

Figure 1.15 Periodic orbits in a discrete-time system

then set $\{\mathbf{x}_0, f(\mathbf{x}_0), \ldots f^{n-1}(\mathbf{x}_0)\}$ is a period-*n* orbit of map *f*. Using a geometrical point of view on dynamical systems, the period orbit of a discrete-time dynamical system is plotted in Figure 1.15.

1.4.2 Linear and Nonlinear Dynamical Systems

Dynamical systems can be further classified according to the equations used to describe the evolution of their state

1. *Linear dynamical system*, if the evolution is governed by a linear differential equation (continuous time) or difference equation (discrete time). Linear dynamical systems can be solved in terms of simple functions, and the behavior of all orbits is classified.
2. *Nonlinear dynamical system*, if the evolution is governed by a nonlinear differential equation (continuous time) or difference equation (discrete time). For example, the aforementioned Lorenz system is a nonlinear continuous-time system.

The theory of linear systems is mature and well developed. Many problems in linear systems theory have been solved using readily available mathematical tools. On the other hand, periodic solutions of nonlinear system are rarely known analytically. But, the study of nonlinear systems is important because almost every real system contains nonlinearities of one form or another [38].

Although there are types of nonlinear systems in mathematics and practical engineering, we will focus on piecewise linear systems because DC–DC converters are naturally modeled as piecewise smooth systems.

1.4.2.1 Piecewise-Smooth Dynamical Systems

The map is *smooth* when the map is continuously differentiable within the whole domain. In mathematics, a piecewise-defined function is defined by multiple sub-functions, and each sub-function applies to a certain interval of the main function's domain (a sub-domain). The map is called *piecewise-smooth* if it has the following two properties: (i) the state-space is divided into different sub-domains, and each sub-domain has independent map expressions and (ii) the map is continuous in the whole state-space, is piecewise continuously differentiable in different sub-domains, and is not differentiable at the boundaries.

The piecewise-smooth dynamical system may switch between different sub-domains, because the system state trajectories may cross one of the state-space boundaries. Undoubtedly, a piecewise-smooth dynamical system is nonlinear dynamical system. There are three classes of piecewise-smooth systems: maps, flows, and hybrid systems [34].

1.4.2.2 Hybrid Dynamical Systems

Hybrid dynamical systems are combinations of maps and flows, giving rise to discontinuous, piecewise-smooth flows. They can arise both as models of impacting systems or in the context of the interaction between digital and analog systems. For example, hybrid automata [35] are defined as dynamical systems with a discrete and a continuous part. The discrete dynamics can be represented as a graph whose vertices are the discrete states (or modes) and whose edges are transitions. The continuous states take values in $\mathbb{R}^n$ and evolve along trajectories, typically governed by ODEs or differential algebraic equations. The interaction between the discrete and the continuous dynamics takes place through invariants and transition relations [34].

For example, a PWM boost converter under peak current control (Figure 1.10) is considered. There are three operation modes, or three topologies, Mode 1, Mode 2, and Mode 3, corresponding to $(S, D) = (\text{on, off})$, $(S, D) = (\text{off, on})$, and $(S, D) = (\text{off, off})$, respectively. Figure 1.16 is the hybrid model of the converter. These topologies are a definitely continuous system and may be described by ordinary differential equations. However, converters are designed to work in these two or three topologies cyclically (CCM or DCM), and the transform from one topology to another is controlled by the external driving signal, which is a typical digital signal or discrete-valued. Therefore, DC–DC converters are definitely piecewise-smooth hybrid systems which include continuous and discrete systems synchronously. Therefore, they may have complex nonlinear behavior.

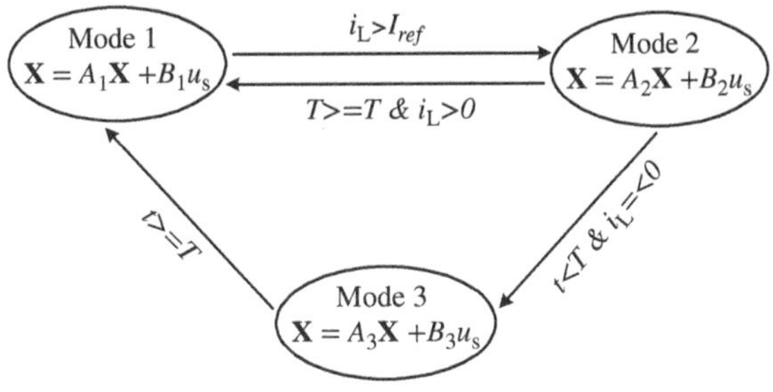

Figure 1.16 Hybrid model of a boost converter in peak-current control

1.4.3 Characterization of Nonlinear Behavior

It is well known that linear methods applied alone cannot give a reasonable explanation of nonlinear phenomena. Thus, some mathematical tools, including phase portraits, bifurcation diagrams, Lyapunov exponents (LEs), power spectrums, Poincaré Sections, Jacobian matrices, spider diagrams, and so on, provide information about the dynamics of the nonlinear dynamical system.

1.4.3.1 Poincaré Map

DC–DC converters are a type of hybrid system managed by a close-loop feedback controller. It is difficult to obtain the analytical solution, even a numerical solution, so some graphical tools are used to analyze these systems. In the previous section, we introduced phase portrait and bifurcation diagrams to describe several motions of a dynamical system. Compared with analytical methods, these methods may simplify complex nonlinear behavior. However, for instance, some periodic motions have high period numbers which perhaps are a mess and difficult to distinguish from aperiodic trajectories. A common mathematical tool called the Poincaré section can solve this question well.

A Poincaré section is a hypersurface Σ in the state-space, which replaces state trajectories of the n-dimensional continuous vector field with an $(n-1)$-dimensional map as shown in Figure 1.17. When this plane has a finite number of fixed points, the original continuous trajectories are periodic. When this plane has a closed curve, the original continuous trajectories are quasi-periodic. When the plane has dense points assembling in a certain zone, the original trajectories are chaotic. It is difficult to obtain the analytical solutions or numerical solutions of the majority of nonlinear continuous-time systems, so the intuitive features of the Poincaré section make the relevant studies convenient.

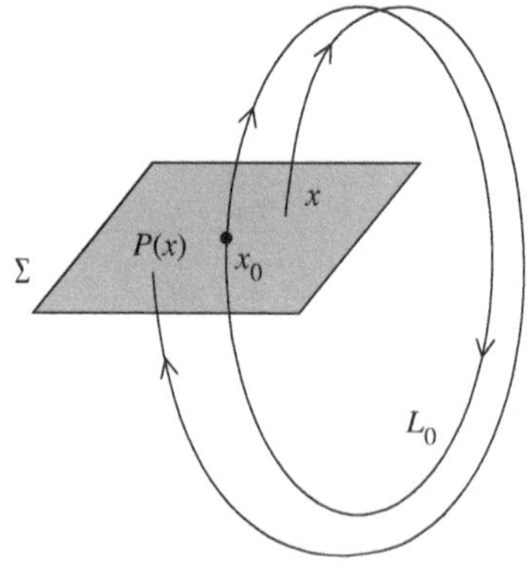

Figure 1.17 Illustration of Poincaré section. With kind permission from Springer Science+Business Media: Applied Mathematical Sciences, "Elements of Applied Bifurcation Theory", 112, ©2000, Kuznetsov, Y. A.

If we observe the representative points at every constant time interval, unlike the continuous motions of the orbits in phase plane, a series of discrete points may be obtained. Studying the motion of the orbits comes down to the motion of discrete points, or a new discrete-time system, or Poincaré map. As already mentioned, the time interval is called the sampling period, the discrete points are called the sampling points, and the new map is called the stroboscopic map. The discretized system thus reveals the periodicity of the orbit of the underlying continuous-time system via the periodicity of the corresponding sampled state.

Next, a buck converter is used as an illustration to describe the chaos motions via the Poincaré map. The buck converter is in voltage control with parameters $L=20\,\text{mH}$, $C=47\,\mu\text{F}$, $A=8.4$, $V_L=3.8\,\text{V}$, $V_U=8.2\,\text{V}$, $T=400\,\mu\text{s}$, $R=22$, and $V_{ref}=11.3\,\text{V}$ as shown in Figure 1.18. The discrete map is modeled according to Equation 1.8, and the Poincaré Section of the inductor current and capacitor voltage at fixed time intervals T is obtained using the discrete map. Figures 1.19a and b show the period-1 and period-2 orbits at $E=22\,\text{V}$ and $E=28\,\text{V}$ respectively. The Poincaré Section with $E=37\,\text{V}$ is shown in Figure 1.19c. The left one is the corresponding phase portrait, and the right is the trajectories plotted as graphs of v and v_{ramp}. From this figure, there are contiguously populated regions, proving that the converter is chaotic.

1.4.3.2 Power Spectrum

Nonrandom motions have a definite frequency structure, and Fourier analysis is a tool to determine the frequency contents of a given signal. If the signal is periodic or quasi-periodic, the power spectrum will consist of a sequence of "spikes" at the fundamental frequencies. If the signal is neither periodic nor quasi-periodic (for example, it is chaotic), then the spectrum will be continuous. Thus, the sudden appearance of a continuous power spectrum from a discrete spectrum is viewed as an indicator of the onset of chaotic behavior when some parameters of the system are changed [40].

A power spectrum is very useful for analysis of a time series, especially for experiment and simulation data. For example, Feigenbaum (1976) used power spectrum

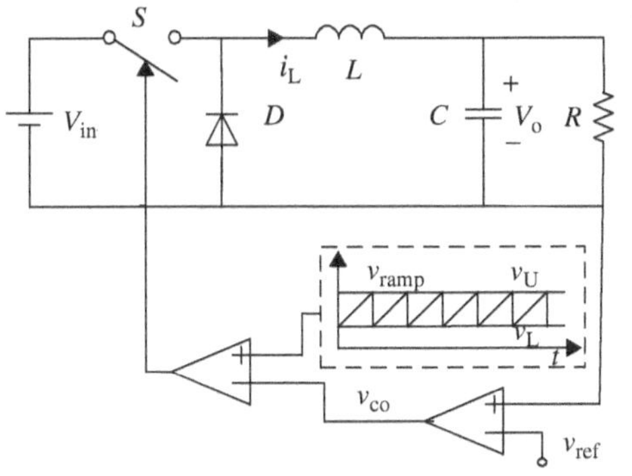

Figure 1.18 Schematic diagram of buck converter

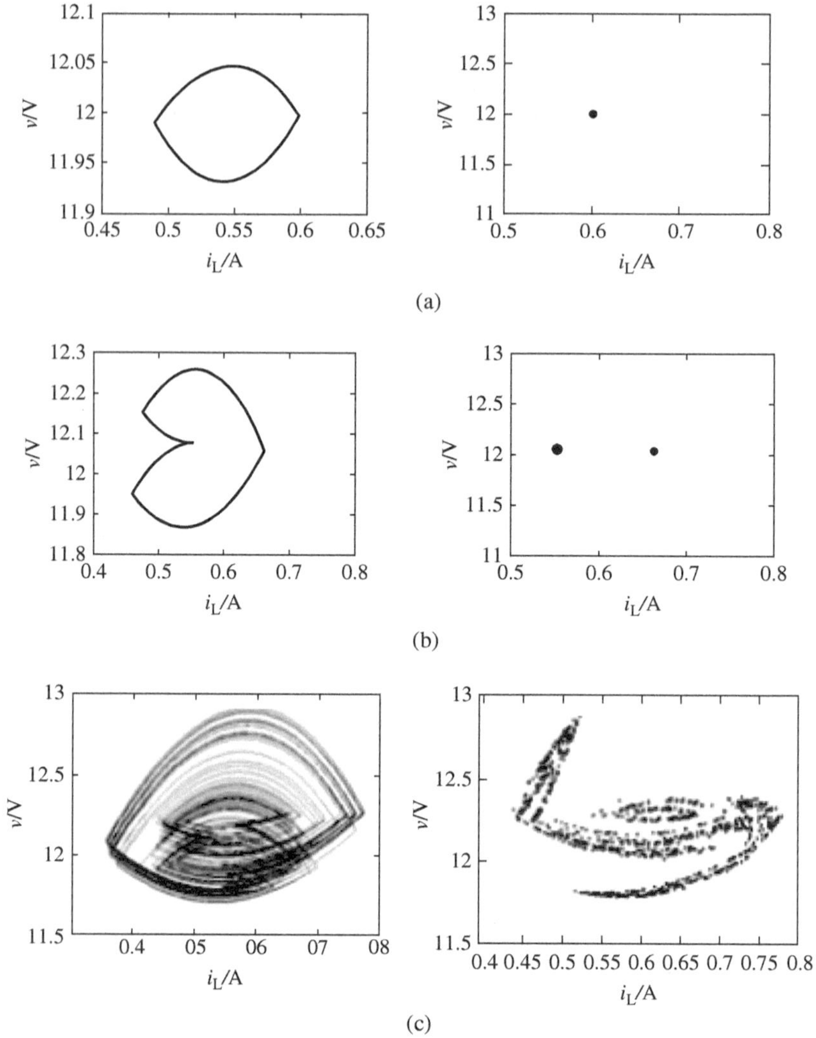

Figure 1.19 Phase portrait and Poincaré section of buck converter for (a) $E = 22$, (b) $E = 28$, and (c) $E = 37$ V

analysis to study the onset spectrum of turbulence [36]. Next, Gollub and Bension used it in a turbulent convection study [37], and it was then used in the study of nonlinear behavior of the Josephson junction by Iansiti *et al.* [41], and so on.

The spectrum structure of period-doubling bifurcation and chaos is illustrated in Figure 1.20, where the higher harmonics are ignored. For period-1 motion, the spectrum is dominated by a primary spectral peak with f_0. For period-2 motion, the subharmonic ($f_0/2$) is excited beside the primary spectral peak. For period-4 motion, the spectrum contains four peaks, corresponding to a new subharmonic ($f_0/4$).

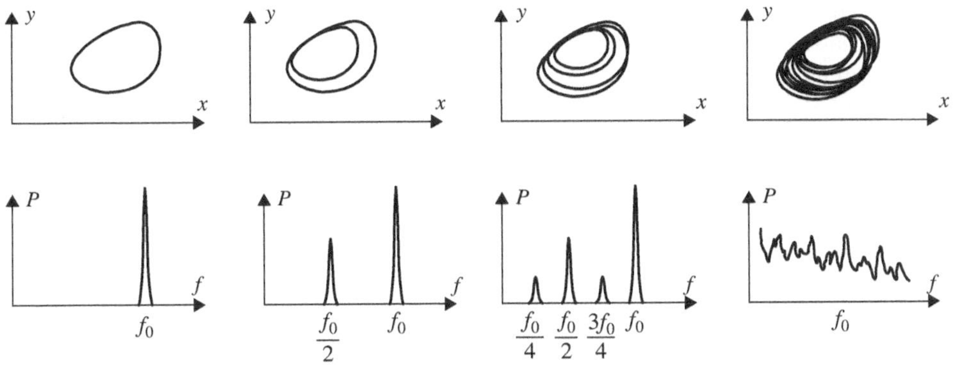

Figure 1.20 Power spectrum of period-doubling bifurcation. From left to right, there are phase trajectories and spectrums of period-1, period-2, period-4, and chaos

When the system becomes chaotic, the spectrum has broadband and continuous characteristics. In other words, chaos motion may be characterized by the presence of broadband and a continuous power spectrum.

1.4.3.3 Jacobian Matrix

The Jacobian matrix is another commonly used mathematic tool aside from the Poincaré map, which is used to analyze the stability of equilibriums or fixed points in continuous or discrete systems.

Suppose that a continuous smooth dynamical system has the form of

$$\begin{cases} \dot{x} = f_1(x, y) \\ \dot{y} = f_2(x, y) \end{cases} \tag{1.13}$$

and x_0, y_0 are its equilibriums. The Jacobian matrix of Equation 1.13 is given by

$$\mathbf{J} = \begin{pmatrix} \frac{\partial f_1}{\partial x} & \frac{\partial f_1}{\partial y} \\ \frac{\partial f_2}{\partial x} & \frac{\partial f_2}{\partial y} \end{pmatrix}\Bigg|_{\substack{x=x_0 \\ y=y_0}} \tag{1.14}$$

Assuming that Matrix **J** has two eigenvalues λ_1 and λ_2, which are the roots of the characteristic equation

$$\lambda^2 - \sigma\lambda + \Delta = 0 \tag{1.15}$$

where $\sigma = \text{tr}(\mathbf{J})$, $\Delta = \det(\mathbf{J})$. The eigenvalues of this matrix determine the stability of the equilibriums. A stable equilibrium attracts nearby trajectories, while an unstable equilibrium repels nearby trajectories. Eigenvalues λ_1 and λ_2 might be real or complex numbers, whose values indicate the classification of equilibrium and fixed points. For detailed classification, please see references [33, 35]. Generally speaking, in the dynamical system, all eigenvalues of the Jacobian matrix will motion on the complex plane with a certain parameter. When the change parameter (called a bifurcation parameter) reaches a critical value, the stability of a periodic orbit changes and bifurcation behavior occurs.

Now we consider a discrete-time dynamical system

$$x \mapsto f(x) \quad x \in \mathbb{R}^n \tag{1.16}$$

where the map f is smooth. Let $x_0 = 0$ be a fixed point of the system (i.e., $f(x_0) = x_0$) and let $\mathbf{J}$ denote the Jacobian matrix df/dx evaluated at x_0. The eigenvalues λ_1, λ_2, …, λ_n of $\mathbf{J}$ are called *multipliers* or *Floquet multipliers* of the fixed point. The absolute values of multipliers may be determined. When all $|\lambda| < 1$, the fixed points are stable. When $|\lambda| > 1$, the fixed points are unstable. Thus, the bifurcation appears at the *Floquet unit circle* ($\lambda = \pm 1$). The standard bifurcation types of discrete system, such as tangent bifurcation, period-doubling bifurcation, and torus bifurcation, may be distinguished by studying the trajectories of the multipliers. For the high dimension map, if all the multipliers are smaller than 1, the fixed points are stable. Conversely, as long as one multiplier is greater than 1, the fixed points are unstable. For power converters, the discrete map has the form

$$\mathbf{x}_{n+1} = f(\mathbf{x}_n, d_n, \mu_n) \tag{1.17}$$

where $\mathbf{x}$ is the state variable, μ_n is the system parameter, d_n is the duty cycle, and the subscript n defines the nth switching period. The stability of the map may be estimated by the characteristic values of the Jacobian matrix. The relationship between d_n and x_n can be expressed using the *switching surface*

$$s(\mathbf{x}_n, d_n) = 0 \tag{1.18}$$

The Jacobian matrix of Equation 1.17 is

$$J(\mathbf{x}^*) = \frac{\partial f}{\partial \mathbf{x}_n} - \frac{\partial f}{\partial d_n}\left(\frac{\partial s}{\partial d_n}\right)^{-1}\frac{\partial s}{\partial \mathbf{x}_n}\Bigg|_{\mathbf{x}_n=\mathbf{x}^*} \tag{1.19}$$

And the eigenvalues are roots of the characteristic equation

$$\det[\lambda I - J(\mathbf{x}^*)] = 0 \tag{1.20}$$

where I is the $n\times n$ identity matrix. When the maximum value of the norms of the characteristic roots λ is greater than 1, Equation 1.17 will not be convergent and the unstable phenomenon will occur.

1.4.3.4 Lyapunov Exponents

Although power spectrum is an intuitive and graphic method, it is not a characteristic quantity of "invariance" in principle. For example, to calculate the power spectrum, the signal $\{x_i\}$ should be changed to $\{x_i^2\}$. Hence, some new frequency components will appear naturally.

LEs are quantitative measures of the evolution of the phase trajectories, and are useful for distinguishing chaos and non-chaos. As we all know, the basic performance of chaos is sensitivity to initial conditions. Eigenvalues of the Jacobian matrix decide the rate of being stretched or expanded of two local neighbor trajectories, but this rate is perhaps different for the other neighbor trajectories in the whole phase space. Therefore, it is very necessary to quantify the overall characteristics of the dynamical system, and LEs are one of the quantity parameters (invariant measures, fractal dimension, etc., are the others). In mathematics, *Lyapunov exponents* are defined as an average rate of exponential divergence of two initially nearby trajectories. Quantitatively, two trajectories in phase space with initial separation δx_0 diverge (provided that the divergence can be treated within the linearized approximation) at a rate given by

$$|\delta x(t)| \approx e^{\lambda t} |\delta x_0| \tag{1.21}$$

where λ is the LE.

What is a geometrical meaning of LEs? The theory is illustrated in Figure 1.21. Consider two points in a phase space, x_0 and $x_0 + \delta x_0$; each of them will generate an orbit using some equations. These orbits can be thought of as parametric functions of a variable like time. Because sensitive dependence can arise only in some portions of a system, this separation is also a function of the location of the initial value and has the form $\delta x(t)$. In a system with attracting fixed points or attracting periodic orbits, $\delta x(t)$ diminishes asymptotically with time. If a system is unstable, the orbits diverge exponentially for a while, but eventually settle down. For chaotic points, the function $\delta x(t)$ will behave erratically. It is thus useful to study the mean exponential rate of divergence of two initially close orbits using the formula

$$\lambda = \lim_{t\to\infty} \lim_{\delta x_0 \to 0} \frac{1}{t} \ln \frac{|\delta x(t)|}{|\delta x_0|} \tag{1.22}$$

This number is called the Lyapunov exponent, and there are three situations:

If $\lambda < 0$, the orbit attracts to a stable fixed point or stable periodic orbit.
If $\lambda = 0$, the orbit is a neutral fixed point (or an eventually fixed point). A LE of zero indicates that the system is in some sort of steady-state mode.
If $\lambda > 0$, the orbit is unstable and chaotic. Nearby points, no matter how close, will diverge to any arbitrary separation.

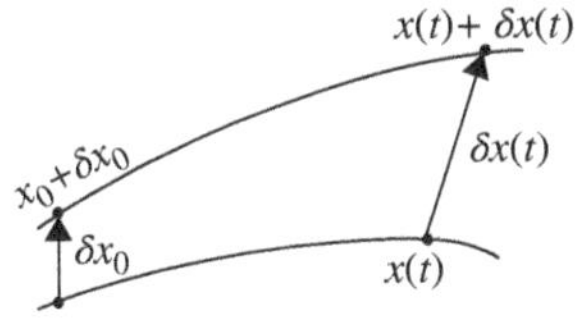

Figure 1.21 Geometrical meaning of Lyapunov exponents

LE is useful to distinguish among the various types of orbits. It works for discrete as well as continuous systems. We can legitimately conclude that the behavior of the system is chaotic only if its average LE is positive [40].

1.5 Conclusions

Switching converters are complex nonlinear, nonautonomous, and time varying systems, so they contain abundant nonlinear phenomena, such as chaos and bifurcation. First, we introduced the working principle and the basic topologies of the switching converters in this chapter. Second, an overview of nonlinear behavior in DC–DC converters was given. Finally, we introduced some reviews of basic nonlinear dynamics concepts, including the dynamical system and some characterizations of nonlinear behavior.

References

[1] Rashid, M.H. (2004) *Power Electronics: Circuits, Devices, and Application*, 3rd edn, Pearson Education.

[2] Agrawal, J.P. (2001) *Power Electronic Systems: Theory and Design*, Prentice-Hall, Upper Saddle River, NJ.

[3] Erickson, R.W. and Maksimovic, D. (2000) *Fundamentals of Power electronics*, 2nd edn, Springer.

[4] Rashid, M.H. (2007) *Power Electronics Handbook*, 2nd edn, Elsevier.

[5] Luo, F.L. and Ye, H. (2004) *Advanced DC/DC Converters*, CRC Press.

[6] Mohan, N., Undeland, T.M. and Robbins, W.P. (1995) *Power Electronics: Converters, Applications, and Design*, 2nd edn, John Wiley & Sons, Inc., New York.

[7] Tse, C.K. (1994) Flip bifurcation and chaos in a three-state Boost switching regulators. *IEEE Trans. Circuits Syst. I*, **41**(1), 16–23.

[8] Tse, C.K. (1994) Chaos from a Buck switching regulator operating in discontinuous mode. *Int. J. Circuits Theory Appl.*, **22**(4), 263–278.

[9] Iu, H.H.C. and Tse, C.K. (2001) Bifurcation behaviour in parallel-connected Buck converters. *IEEE Trans. Circuit Syst. I*, **48**(2), 233–240.

[10] Ma, Y., Tse, C.K., Kousaka, T. and Kawakami, H. (2005) Connecting border collision with saddle-node bifurcation in switched dynamical systems. *IEEE Trans. Circuit Syst. II*, **52**(9), 581–585.

[11] Tse, C.K., Lai, Y.M. and Iu, H.H.C. (2000) Hopf bifurcation and chaos in a free-running current-controlled Ćuk switching regulator. *IEEE Trans. Circuit Syst. I*, **47**(4), 448–457.

[12] Aroudi, A.E., Benadero, L., Toribio, E. *et al.* (1999) Hopf bifurcation and chaos from torus breakdown in a PWM voltage-controlled DC-DC boost converter. *IEEE Trans. Circuit Syst. I*, **46**(11), 1374–1382.

[13] Iu, H.H.C. and Tse, C.K. (2003) Study of low-frequency bifurcation phenomena of a parallel-connected boost converter system via simple averaged models. *IEEE Trans. Circuit Syst. I*, **50**(5), 679–685.

[14] Dai, D., Tse, C.K. and Ma, X. (2005) Symbolic analysis of switching systems: application to bifurcation analysis of DC-DC switching converters. *IEEE Trans. Circuits Syst. I Regul. Pap.*, **52**(8), 1632–1643.

[15] Banerjee, S. (1997) Coexisting attractors, chaotic saddles, and fractal basins in a power electronic circuit. *IEEE Trans. Circuits Syst. I*, **44**(9), 847–849.

[16] Wong, S.C., Tse, C.K. and Tam, K.C. (2004) Spurious modulation on current-mode controlled DC/DC converters: an explanation for intermittent chaotic operation. *IEEE ISCAS*, **5**, 852–855.
[17] Tse, C.K., Zhou, Y., Lau, F.C.M. and Qiu, S. (2003) "Intermittent" chaos and subharmonics in switching power supplies. *IEEE ISCAS*, **3**, 332–335.
[18] Tse, C.K. (2004) *Complex Behavior of Switching Power Converters*, CRC Press.
[19] Aston, P.J., Deane, J.H.B., anHamill, D.C. (1997) Targeting in systems with discontinuities, with applications to power electronics. *IEEE Trans. Circuits Syst. I*, **44**(10), 1034–1039.
[20] Deane, J.H.B. and Hamill, D.C. (1996) Improvement of power supply EMC by chaos. *Electron. Lett.*, **32**, 1045–1045.
[21] Morel, C., Bourcerie, M. and Chapeau-Blondeau, F. (2004) Extension of chaos anticontrol applied to the improvement of switch-mode power supply electromagnetic compatibility. *IEEE Int. Symp. Ind. Electron.*, **1**, 447–452.
[22] Zhang, J.J., Zhang, L.M. and Qiu, S.S. (2006) An experimental investigation of EMI suppression of off-line switching converter by chaotic modulation. *Aerosp. Control*, **24**(4), 87–90.
[23] Tse, K.K., Chung, H.S.-H., Huo, S.Y. and So, H.C. (2000) Analysis and spectral characteristics of a spread-spectrum technique for conducted EMI suppression. *IEEE Trans. Power Electron.*, **15**(2), 399–410.
[24] Tse, K.K., Chung, H.S.-H., Huo, S.Y. and So, H.C. (2002) A comparative study of carrier- frequency modulation techniques for conducted EMI suppression in PWM converters. *IEEE Trans. Ind. Electron.*, **49**(3), 618–627.
[25] Tse, K.K., Ng, R.W.-M., Chung, H.S.H. *et al.* (2003) An evaluation of the spectral characteristics of switching converters with chaotic carrier-frequency modulation. *IEEE Trans. Ind. Electron.*, **50**(1), 171–182.
[26] Wang, X.M. (2008) Symbolic time series characterization and block entropy analysis of DC-DC converters. *Acta Physiol. Scand.*, **57**(10), 6112–6119.
[27] Deane, J.H.B. (1992) Chaos in a current-mode controlled dc-dc converter. *IEEE Trans. Circuits Syst. I*, **39**(8), 680–683.
[28] Banerjee, S. and Verghese, G.C. (2001) *Nonlinear Phenomena in Power Electronics: Attractors, Bifurcations, Chaos, and Nonlinear Control*, Wiley Press.
[29] Brockeet, R.W. and Wood, J.R. (1984) Understanding power converter chaotic behaviour mechanisms in protective and abnormal modes. 11th Annual International Power Electronics, pp. 115–124.
[30] Hamill, D.C. and Jefferies, D.J. (1988) Subharmonics and chaos in a controlled switched-mode power converter. *IEEE Trans. Circuits Syst. I*, **35**(8), 1059–1061.
[31] Kuznetsov, Y.A. (2000) *Elements of Applied Bifurcation Theory*, 2nd edn, Springer-Verlag.
[32] Wang, X.M., Zhang, B., and Qiu, D.Y. (2008) Bifurcations and chaos in H-bridge DC chopper under peak-current control. International Conference on Electrical Machines and Systems, 2008. ICEMS 2008, Wuhan, October 17–20, 2008, pp. 2173–2177.
[33] Nayfeh, A.H. and Balachandran, B. (2004) *Applied Nonlinear Dynamics*, Wiley Press.
[34] Bernardo, M., Budd, C.J., Champneys, A.R. *et al.* (2007) *Piecewise-smooth Dynamical Systems: Theory and Applications*, Springer-Verlag Press.
[35] Branicky, M.S., Borkar, V.S. and Mitter, S.K. (1998) A unified framework for hybrid control model and optimal control theory. *IEEE Trans. Autom. Control*, **43**(1), 31–45.
[36] Feigenbaum, M.J. (1980) The metric universal properties of period doubling bifurcations and the spectrum for a route to turbulence. *Ann. New York. Acad. Sci*, **357**, 330–336.
[37] Gollub, J.P. and Benson, S.V. (1980) Many routes to turbulent convection. *J. Fluid Mech.*, **100**(3), 449–470.
[38] Liu, D. and Michel, A.N. (1994) *Dynamical Systems with Saturation Nonlinearities*, Springer-Verlag Press.

[39] Wang, X.M. (2009) *Study and Application of Bifurcation and Chaos Behavior of Power Electronic Converters Based on Discrete System Dynamics(in Chinese)*, South China University of Technology, Guangzhou.

[40] Hilborn, R.C. (2000) *Chaos and Nonlinear Dynamics: An Introduction for Scientists and Engineers*, 2nd edn, Oxford University Press.

[41] Iansiti, M., Hu, Q. and Westervelt, R.M. (1985) Noise and chaos in a fractal basin boundary regime of a Josephson junction. *Phys. Rev. Lett.*, **55**(7), 746–749.

[42] Huang, R.S. and Huang, H. (2005) *Chaos and its Application (in Chinese)*, 2nd edn, Wuhan University Press.

2

Symbolic Analysis of the Nonlinear Behavior of DC–DC Converters

2.1 Introduction

The different types of bifurcation is still one of the main areas of research in the nonlinear behavior of switching power converters. Thus far, the further study of converters' bifurcation and chaos behavior has had two challenges: first it is difficult to obtain the necessary analytic estimates because of switching nonlinearities; second, chaos measurable properties, such as power spectrums, Lyapunov exponents, fractal dimensions, and entropy, are too complex for engineering applications.

In recent years, a new symbolic method has been proposed for analysis of bifurcations in switching power converters. In Dai and Tse [1–3], the switching states are encoded as the binary symbols "0" or "1," and two new concepts – block sequence and block number – are defined and used to detect both border collision and standard bifurcations. This symbolic analysis method of the above studies reveals the mechanism of bifurcations from a circuit operational viewpoint, leading to a direct detection criterion for global bifurcation, and takes the initiative to apply symbolic methods to the nonlinearity studies of switching power converters. In this chapter, our intention is to enrich the existing studies on symbolic analysis, and to develop simpler methods to obtain the characterization of nonlinearities of switching power converters. The nonlinear behavior of switching power converters from a statistics perspective, instead of a circuit operational topologies perspective, is quantified.

Additionally, it may be meaningful to choose a very coarse-grained description. Hence symbolic dynamics and symbolic time series analysis, which are powerful techniques in describing dynamical behavior by symbolizing a data series and have been successfully used in areas including communication, mechanics, medical science, electrochemistry, and cryptology [4–8]. In [9], the symbolic dynamics analysis has the merits of: high efficiency numerical analysis, strong real-time, low cost and low complexity, and having no sensitivity to the measurement noise. Therefore, symbolic

Chaos Analysis and Chaotic EMI Suppression of DC-DC Converters, First Edition. Bo Zhang and Xuemei Wang.

dynamics is used to analyze the nonlinearities of switching power converters in this chapter [10, 11].

This chapter has been structured as follows. In Section 2.2 period-doubling cascades and the symbolic dynamics method are introduced. In Section 2.3 we analyze the bifurcation and chaos behavior of two boost converters via Shannon entropy, by which both period-doubling bifurcation and border collision bifurcation are verified and quantified. Initial value sensitivity is successfully used to distinguish chaos from periodic motions. The results show that this symbolic method strongly correlates with the Lyapunov exponent and can be calculated more easily and faster than the Lyapunov exponent. We think this exercise opens up possibilities in the study of nonlinear phenomena by symbolic dynamics and can be applied directly to other switching power converters.

2.2 Overview of the Time Series Principle of Discrete Systems

2.2.1 *Symbolic Dynamics and Symbolic Time Series*

Symbolic dynamics, an algebraic approach, originated from the abstract topological theory of dynamic systems and developed in the 1930s, gradually becoming a powerful technique for analysis of dynamical behavior with symbols. Research of chaos phenomenon became increasingly active from the 1960s. Evolving from abstract symbolic dynamics, *Practical Symbolic Dynamics* became a useful and clever tool for practical researchers.

Symbolic dynamics is based on a coarse-graining of the measurements. For example, the data sampled every t interval are transformed into a pattern whose elements are only a few letters from some alphabet. This way, the study of the dynamics simplifies to the description of symbol sequences. In doing so, one loses some of detailed information, but some of the invariant, robust properties of the dynamics may be kept [7]. Symbolic dynamics provides the most rigorous means to describe dynamic behavior with finite precision. Symbolic dynamics is applicable in the high dimension dynamic system, but in a low dimension it may develop rich content. So, in this section, the symbolic dynamics theory of the unimodal map is introduced first.

Consider an iterative function

$$x_{n+1} = f(x_n, \mu) \tag{2.1}$$

where μ is the given parameters. By iterating Equation 2.1 and giving explicitly the function f and an initial value x_0, one can compute a sequence

$$x_0,\ x_1 = f(x_0),\ x_2 = f(x_1),\ \cdots, x_n = f(x_{n-1}),\ \cdots \tag{2.2}$$

Usually, the parameter μ and function f are chosen and normalized in such a way that the iterate x_{n+1} belongs to the same interval if x_n is taken from a finite interval I. Therefore, the function f maps the interval I into itself, whence followed the name "map of the interval." We say "into itself," because, even when x_n takes all possible values from I, x_{n+1} may not fill up the entire interval I [12].

In order to demonstrate the usefulness of symbolic description, we will introduce this process in this simple case of the logistic map

$$x_{n+1} = \mu x_n(1 - x_n) \tag{2.3}$$

where μ is the parameter. This map has two monotonic branches which are separated by a critical point C at the maximum of the map with coordinate x_c (Figure 2.1). In symbol dynamics, the discrete data series Equation 2.2 should be initially converted into a symbolic series by introducing a partition of the phase space and assigning a symbol to each cell of the partition [13]. Next, we only consider where x_n is located, to the left or the right of the critical point C which separates the interval into two segments based on the monotonic branches of the map. Each segment needs a letter for a label, for example, two letters R and L, like

$$s_n = \begin{cases} L, & \text{if } x_n < x_C \\ C, & \text{if } x_n = x_C \\ R, & \text{if } x_n > x_C \end{cases} \tag{2.4}$$

where x_C is the coordinate of the critical point C. Such a map is called unimodal because a single critical point is involved. One can form infinitely symbolic sequences via continuous iterations of Equation 2.3 with a given initial x_0

$$s(x_0) = s_0 s_1 s_2 \cdots s_n \cdots \tag{2.5}$$

Today, symbolic sequences of three letters for unimodal mappings have been constructed.

Obviously, the description of symbolic series Equation 2.5 is coarser than data series Equation 2.2. However, this coarse and simplified description method can reflect some

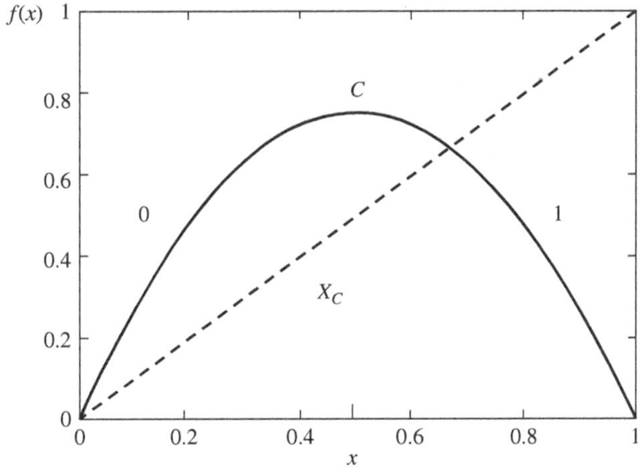

Figure 2.1 Partition of logistic map induced by critical point C ($\mu = 3$). Here, the diagonal is the iteration line $x = f(x)$

of the nature of nonlinear behavior better for the following reasons. First, there is the possible relationship of many-to-one, because many data orbits convert one symbolic series. Thus, this relationship may provide a classification method for different orbits. Symbol sequences reflect only the characteristics that the system is stretching and folding. Moreover, monotonicity is considered regardless of the specific form of the mapping. Therefore, the results obtained by the analysis of symbolic sequences may be universal, easy to describe and analyze. And they could provide a scientific classification method for the dynamic system.

Time series implies a rich system dynamics; so how to extract this information and apply it to practical engineering is an important aspect of chaos application. Studies have shown that symbolic dynamics has become a powerful tool in the field of system recognition in physics and engineering. Table 2.1 illustrates the application of symbolic dynamics in the different disciplines [14]. Engineering applications and theoretical analysis of symbolic dynamics have different focuses. In engineering, symbolic dynamics is used to extract the useful system information as topology information. Furthermore, the encoding method of symbolic dynamics provides a relatively simple, strongly robust, and effective method in system parameter identification.

2.2.2 *Symbolization Method*

How to symbolize discrete time series is the first problem to be solved before using symbolic dynamics. In practice, a suitable coarse description method must reflect geometric or physical characteristics of the dynamic process. The idea behind threshold crossings is natural and fundamental and based on a physical model. In science, researchers wishes to interpret these collecting data. This may mean associating adjectives, that is, "hot" versus "cold." In such cases, it is necessary to partition these states. On the other hand, one may wish to retain pure measurements [15].

Of course, different symbolization methods will make the time sequence change significantly, so it is important to choose a suitable symbolization method. Symbolization methods commonly used in time series include the following methods: the averaged method, one order difference method, and partition method.

Table 2.1 Symbolic dynamics in mathematics, physics, and engineering

Disciplines	Method	Object	Emphasis
Mathematics	Generation partition in phase space	1-1, n-1 mapping	Proving
Physics	Approximate generation partition by experiment results	System model, state identification, verification	Understanding
Engineering	Data compression, simplification, and robustness	Diagnosis, identification	Practice

2.2.2.1 Averaged Method

The process is illustrated in Figure 2.2, and the points in the figure are the discrete samplings of time data series. An average value of a time series x_{AVE} is defined as

$$x_{\mathrm{AVE}} = \frac{1}{n}\sum_{i=1}^{n} x_i \tag{2.6}$$

Then the symbolic series may be established as

$$s_i = \begin{cases} 1, & \text{if } x_i > x_{\mathrm{AVE}} \\ 0, & \text{if } x_i \le x_{\mathrm{AVE}} \end{cases}$$

This classification method is not sensitive to noise because only the overall characteristics, not the local characteristics, of sequences may be expressed. Meanwhile, many details are erased. The one order difference method may be utilized to give more expression of details.

2.2.2.2 One Order Difference Method

Compared with the front data, if this point is larger than the front one, the symbol is taken as "1," otherwise it is "0" in Figure 2.3. The symbol sequence "01" or "10" corresponds to a local extremum. The detail component of the signal is largely extracted in this method.

2.2.2.3 Partition Method

For an obvious structure system, each symbol can be obtained from a given domain of the phase space based on its topological properties. For example, in a logistic map system, the partition is located at the differentiable maximum which separates the two monotonic branches. So the increasing and decreasing branches correspond to the symbol "0" and "1," respectively [16]. This partition method is utilized for systems

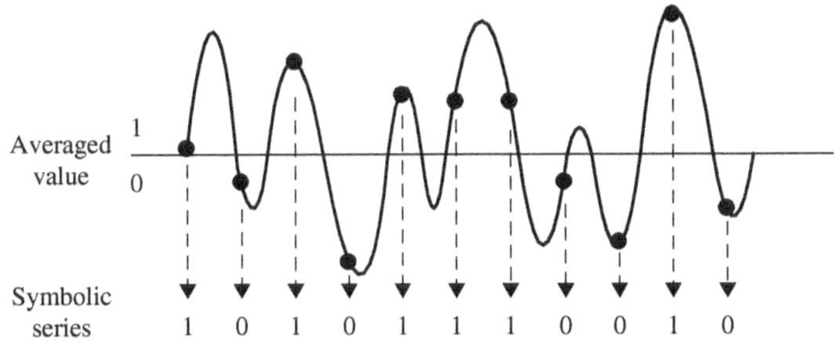

Figure 2.2 Schematic diagram of averaged partition

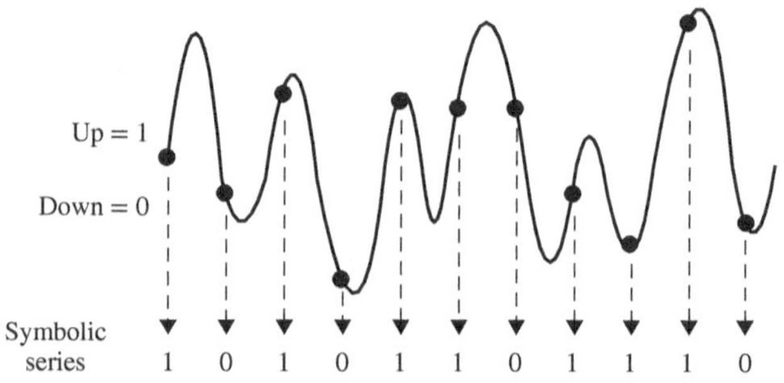

Figure 2.3 Schematic diagram of first order difference

that have obvious structure characteristics, such as some natural extreme points existing in neurobiology and chemistry [7, 17].

Unfortunately, a good partition is difficult to find in the foregoing literatures when the researchers generate the partitions in nonhyperbolic dynamical systems [15]. Different kinds of partition methods have been used in various fields of scientific research. *Statistical partition* (or equal-probable partition), using a partition leading to equal-probable symbols, has been utilized by Tang [13, 18] and Goddler and Letellier [6] to analyze free liquid jets in order to distinguish random and nonrandom behavior. *Approximated partition* based on the maximum relative number of sequences is recorded on the copper electrodissolution, and provides a partition closer to the topological partition [16]. The partition is called *Generation Partition* as a symbol sequence uniquely corresponds to one determinate point on the phase space. In another words, the symbolic series developed from generation partition is unique.

2.2.3 *Symbolic Dynamics of a Period-Doubling Cascade*

Returning to the symbolic series studies, two problems arise naturally. First, how to find the law governing the appearance of the regular regimes? Second, how to characterize the irregular orbits?

The period-doubling bifurcation is the most common bifurcation behavior of switching power converters. In order to demonstrate the usefulness of symbolic description, we shall study the symbolic dynamics of periodic orbits of one-dimensional mappings first. Figure 2.4 is the fixed points diagram of a logistic map with various μ, in which x_p is a solution of

$$x^* = f(x^*, \mu) \tag{2.7}$$

One says that the mapping has reached a *fixed point* (namely period-1) where the line and map curve intersect.

With an increasing parameter μ, the fixed point x_p may move from the increasing branch to the decreasing branch. The fixed points are then encoded as follows

$$(0)^\infty \to (C)^\infty \to (1)^\infty$$

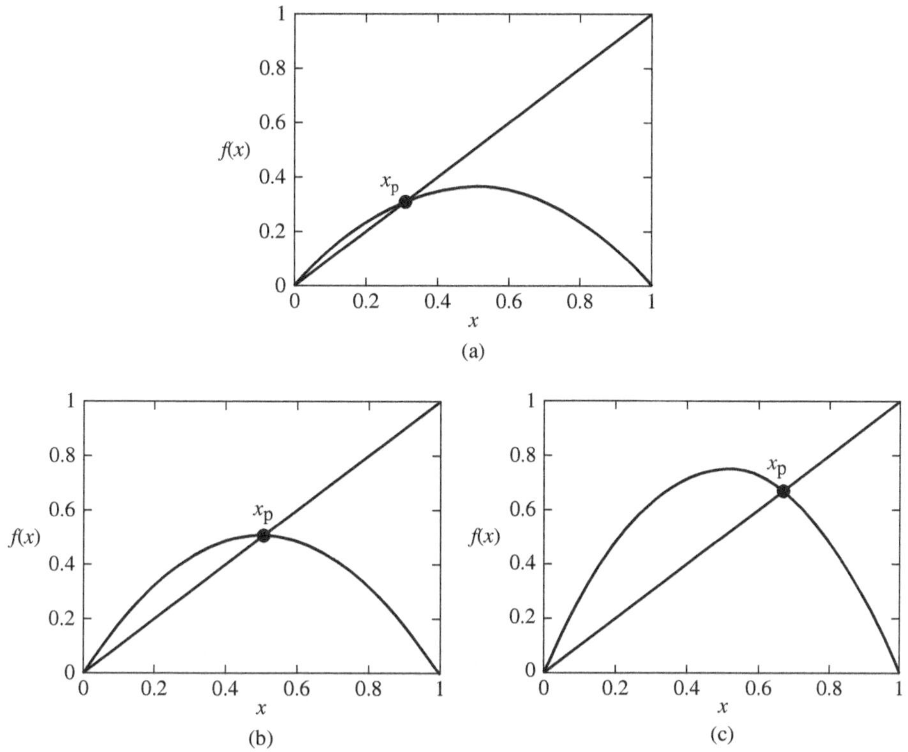

Figure 2.4 Fixed points of logistic map. The diagonal is the iterative line of $x^* = f(x^*, \mu)$ at (a) $\mu = 1.5$, (b) $\mu = 2$, and (c) $\mu = 3$

where $(\mathrm{C})^\infty$ is the a special fixed point, at which the derivative of $f(x)$ is vanishing, and the orbit is called *superstable*. Meanwhile, the fixed point is called the period-1 orbit.

Next, we will briefly describe how a period-doubling cascade is encoded starting from the period-1 orbit. If the slope of the fixed point x_p is less than -1 (Figure 2.4c), this indicates that the fixed point is unstable in the decreasing branch, and the period-2 orbit emerges. Therefore, the period-doubling bifurcation will definitely fall into the map's decreasing branch. So, the period-2 orbit must be encoded as two-bit codes beginning with character "1" as parameter μ varies

$$(11)^\infty \to (1\mathrm{C})^\infty \to (10)^\infty$$

Based on the theory of symbolic dynamics, three symbol replacement schemes of self-similarity can be performed continuously as follows

$$\begin{cases} 1 \to 10 \\ 0 \to 11 \\ \mathrm{C} \to 1\mathrm{C} \end{cases} \tag{2.8}$$

Subsequently, the period-4 orbit is generated after two period-doubling bifurcations. Thus, the symbolic representation of period-4 is

$$(1010)^{\infty} \to (101C)^{\infty} \to (1011)^{\infty}$$

Continuously carrying out the code replacing, encodings of the whole period-doubling cascades are obtained [19]:

$$(0)^{\infty} \to (C)^{\infty} \to (1)^{\infty}\text{Period-1}$$

$$\downarrow$$

$$(10)^{\infty} \leftarrow (1C)^{\infty} \leftarrow (11)^{\infty}\text{Period-2}$$

$$\downarrow$$

$$(1010)^{\infty} \to (101C)^{\infty} \to (1011)^{\infty}\text{Period-4}$$

$$\downarrow$$

$$(10111011)^{\infty} \leftarrow (1011101C)^{\infty} \leftarrow (10111010)^{\infty}\text{Period-8}$$

$$\cdots\cdots$$

For the sake of simplicity, symbol C may be replaced by "0" or "1" arbitrarily in the following, and encodings of the period-doubling cascade have the following rules:

1. Every period may be expressed as two kinds of symbolic code, such as period-2 encodes as $(11)^{\infty}$ or $(10)^{\infty}$, period-4 encodes as $(1010)^{\infty}$ or $(1011)^{\infty}$.
2. Symbolic dynamics may not be used to detect the locations of period-doubling bifurcations, because the codes only change by passing the superstable period, as period-2 transfers from $(11)^{\infty}$ to $(10)^{\infty}$, period-4 transfers from $(1010)^{\infty}$ to $(1011)^{\infty}$, and so on. In other words, the codes will not change by passing the bifurcation points.
3. Symbolic codes, having an odd number of "1," may then induce a period-doubling cascade. For example, the period-4 code $(1011)^{\infty}$ has an odd number of 1, and may induce the next period-8 bifurcation.

2.3 Block Entropy

A symbolic series retains much of the important temporal information of a dynamical system. Characteristics of a symbolic series should be carefully evaluated after partitions are selected. It is possible to introduce estimators of probability which refer to the symbolic statistics of the data. Figure 2.5 illustrates this process. In this example, the symbol sequences are constructed from the three successive binary symbol values occurring at each point in time. Each possible sequence is represented by its binary number equivalent (or the decimal value) determined by the position of each symbol in the template [9]. We can make a conclusion: after symbolization, a moving-frame,

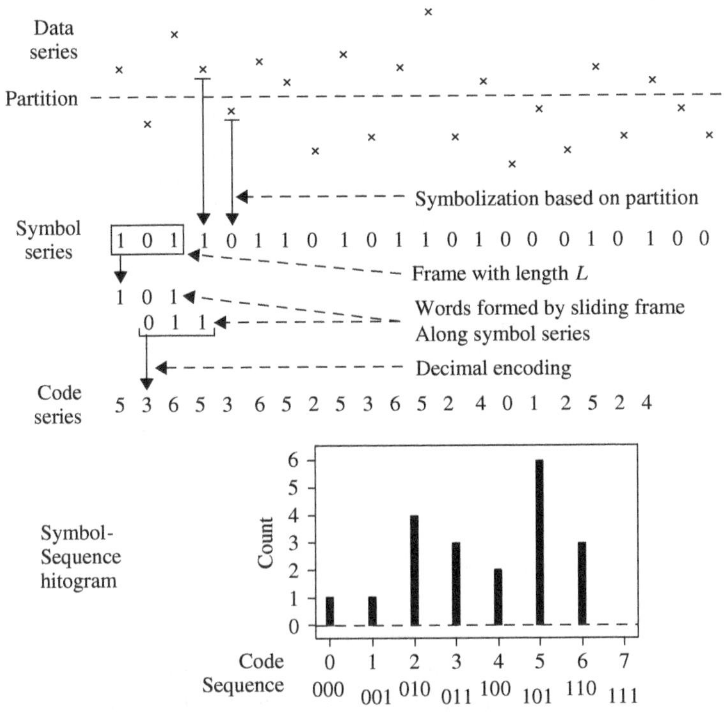

Figure 2.5 Schematic diagram of time series symbolization [9]. Reprinted with permission from Daw, C.S., Finney, C., Tracy E.R. A review of symbolic analysis of experimental data. Rev. Scientific Instruments, 74(2), 915–930. Copyright [2003], AIP Publishing LLC

with suitable length *L*, slides along the symbolic series. Every frame (called subsequence) is then translated into decimal encodings and the probabilities of various symbol subsequences are then measured.

Let p_i be the probability of a subsequence in a symbolic series

$$p_i = \frac{\text{No. of subsequence } i}{\text{Total no. of subsequences}} \tag{2.9}$$

Obviously, the probability of subsequence "5" is 6/20 in Figure 2.5.

In statistics, a histogram is a graphical representation showing a visual impression of the distribution of data. A histogram consists of tabular frequencies and an area equals the frequency of the observations in the interval. The height of a rectangle is also equal to the frequency density of the interval. The total area of the histogram is equal to the number of data. Histograms are used to plot density of data, and are often used for density estimation by estimating the probability density function of the underlying variable [20]. Now, a histogram of a symbol series of logistic map is schemed in Figure 2.6, where length $L = 8$ with two symbols "0" and "1," and the number of subsequences calculated is 2000.

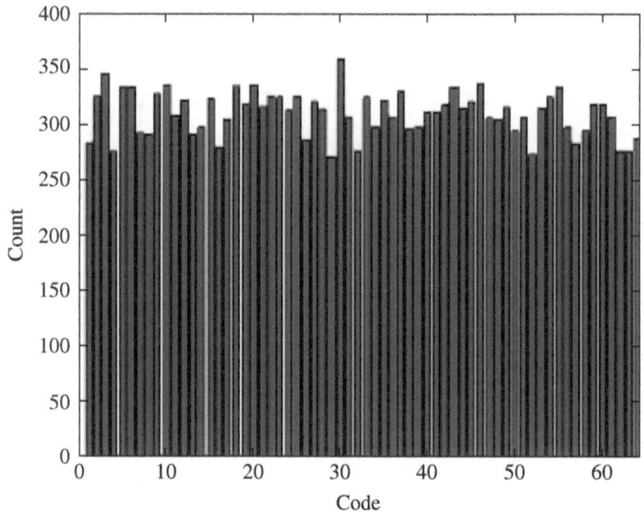

Figure 2.6 Symbol series histogram of logistic map with $L = 8$

In order to quantify the nonlinear behavior of the dynamic system, a concept termed *Shannon entropy* from information theory is illustrated to measure the uncertainty of a discrete finite set, because the histogram is a graphical representation. In fact, entropy is a concept derived from thermodynamics, and was extended to the field of chaotic dynamics by Shanon and Kolmogorov in 1970s. Moreover, some generalized entropy as indices of nonlinearity of the inspection system also has a very good effect. Kolmogorov entropy can be used to distinguish whether the system is regular, random, or chaotic. It is one of the most commonly used dynamical indices in nonlinear dynamics. However, there are two disadvantages in practical application. First, the required amount of data are enormous even in a low-dimensional chaotic system, and at least tens of thousands of sample data are required to be satisfactory. Second, it is particularly sensitive to noise and lets calculations be divergent.

Shannon entropy in information theory is used to measure the uncertainty of a set of discrete finite random events. Shannon entropy is a standard to simultaneously measure the information's ordering and complexity. When the signal is regular and orderly, the corresponding entropy value is small. In contrast, the corresponding entropy value is large when the signal is disorganized. With the evolution of a chaotic system, the regions near the chaotic trajectory have exponential separation take place. Thus, the trajectory cannot be distinguished in a periodic state, but it may be distinguished in a chaotic state. This process is called "chaos creates information."

An average uncertainty is quantified by Shannon's entropy and is defined as

$$H_{\mathrm{L}} = -\sum_{i=1}^{N} p_i \log_2 p_i \tag{2.10}$$

where p_{i} is the probability of symbolic subsequences, N is the number of the symbolic sequence, and $N = 2^{\mathrm{L}}$. Specifically, when $p_{\mathrm{i}} = 0$, assigns $0 \cdot \log_2 0 = 0$.

Shannon entropy gives the quantitative characterizations of a symbolic sequence in [7, 22]:

(i) The entropy value is zero for constant symbolic series.
(ii) For uniform distributions, it takes its maximum value L.
(iii) The entropy value increases with growing L.
(iv) $H_{\mathrm{L(periodic)}} = \log_2 m$, when the system is periodic and the period number $m < L$.

Proof:

(i) The whole sequence is only one symbol for a constant symbolic series, which means that the probability of this symbol is 1 and the probabilities of the other symbols are all 0. Therefore, the entropy value of the constant symbolic series is 0.
(ii) According to information theory, the information entropy has an extreme value when each event has equal probability. Here, the extremum is $H_{\mathrm{S}}(p_1, p_2, \ldots p_N) \leq H_{\mathrm{S}}\left(\frac{1}{N}, \frac{1}{N}, \ldots \frac{1}{N}\right) = \log N$. If the length of the binary sequence is L, the maximum value of entropy is $H_{\mathrm{BS(max)}} = \log_2 2^L = L$. Meanwhile, from an information theory point of view, the dynamical system is full of randomness.
(iii) Because the characteristic of the logarithmic is monotonically increasing and the maximum value of entropy is L, the greater the number of symbolic subsequences, the larger the entropy value.
(iv) $p_1 = p_2 = \ldots = p_{\mathrm{m}} = 1/\mathrm{m}$ when the symbolic dynamical system is periodic, so $H_{\mathrm{BS}}(periodic) = \log_2 m$.

We can estimate whether the time series are periodic or not through (i) and (iv) and whether it is random or pseudorandom through (ii). If its entropy value equals L, it is completely random.

Using Equation 2.10, the block entropy scheme of a logistic map is shown in Figure 2.7. The period-doubling bifurcations can be observed clearly. Obviously, if the period is larger than 8, they cannot be distinguished because the length L of a subsequence is 8. The following conclusions can be deduced from this figure.

1. When the entropy value keeps constant (horizontal lines in Figure 2.7), the system is periodic.
2. The entropy value increases with the growth of chaos behavior.
3. The entropy value is less than L.

The symbol codes, the block entropy of period-doubling cascades, and the tangent bifurcations are tabulated in Table 2.2. To note here, the value of the block entropy depends on the word length L and the distribution of p_{i}, so the entropy value is closely related to the symbolization methods and accuracy. The choice of a specific word length is commonly based on experience. Meanwhile, information is perhaps missed when the word length is too short. In a specific study, the choice of word length may

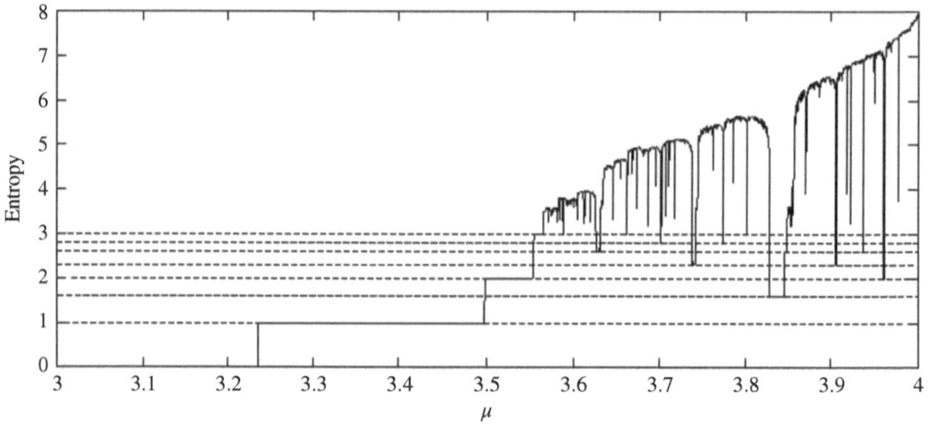

Figure 2.7 Shannon entropy of logistic map. The dashed lines overlap with the entropy curve, the overlap represents period-2, 3, 4, 5, 6, 7, and 8 from bottom to top

Table 2.2 Shannon entropy analysis

	Period number	Symbol codes	H
Period-doubling bifurcation	Period-1	$(0)^{\infty}$, $(1)^{\infty}$	0
	Period-2	$(10)^{\infty}$	1
	Period-4	$(1011)^{\infty}$	2
	Period-8	$(10111010)^{\infty}$	3
Tangent bifurcation	Period-3	$(101)^{\infty}$	1.585
	Period-5	$(10110)^{\infty}$	2.322
	Period-6	$(101110)^{\infty}$	2.585
	Period-7	$(1011011)^{\infty}$	2.807
Chaos		No certain symbolic sequences	Uncertain

be based on experience, too short an L may miss information and too long an L may introduce noise.

Research shows that when L is greater than a certain value, the entropy change is not saturated. From a statistical point of view, entropy is just used to distinguish the complexity of the time process and to characterize the differences or changes of the dynamic systems, rather than to describe the whole characters of the attractors. Thus, only the short data can be used to estimate the entropy.

2.4 Symbolic Time Series Analysis of DC–DC Converters

Our approach is to symbolize the time series based on the iteration map of the converters, and then to calculate the Shannon entropy to analyze its bifurcation and chaos behavior.

2.4.1 Period-Doubling Bifurcation and Chaos of DC–DC Converters

A boost converter with voltage feedback control is shown in Figure 2.8, where X is a desired stable output voltage, D is a desired duty cycle of steady state, x is output voltage, and k is the proportion gain. The parameters are chosen as follows. $E = 16$ V, $L = 208$ μH, $R = 12.5\ \Omega$, $C = 222$ μF, $X = 25$ V, $D = 0.2874$, and $f = 3$ kHz. The corresponding iteration map has been approximately derived in Tse [23]. Figure 2.9 is a time-domain waveform with gain $k = 0.16$. The dots are discrete sampling values at the edge of each clock pulse. Obviously, waveforms are chaos and the saturated duty cycle appears frequently. Then, the gain k is taken as a bifurcation parameter changing from 0.07 to 0.16, and a bifurcation diagram can be generated as shown in Figure 2.10. Period-doubling bifurcations, tangent bifurcations, and chaos can be observed in this diagram.

The return maps are shown in Figure 2.11. The maps are inverse unimodal maps with a critical point $C \approx 26.4$. In order to obtain the symbolic series, we define

$$s_n = \begin{cases} 1, & \text{if } x_n < 26.4 \\ 0, & \text{if } x_n \geq 26.4 \end{cases}$$

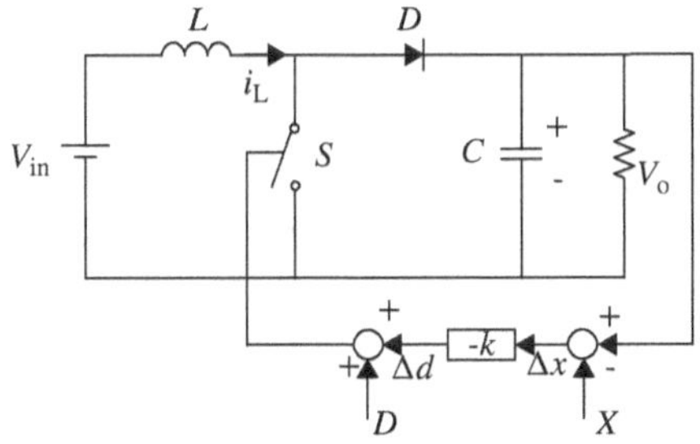

Figure 2.8 Boost converter with voltage feedback control

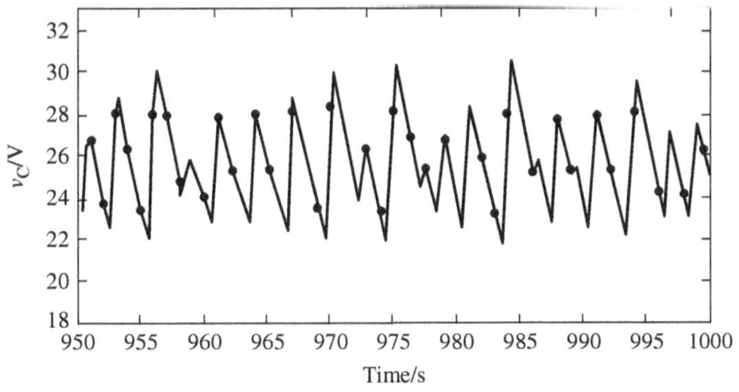

Figure 2.9 Time-domain waveform and discrete data sampled (dots)

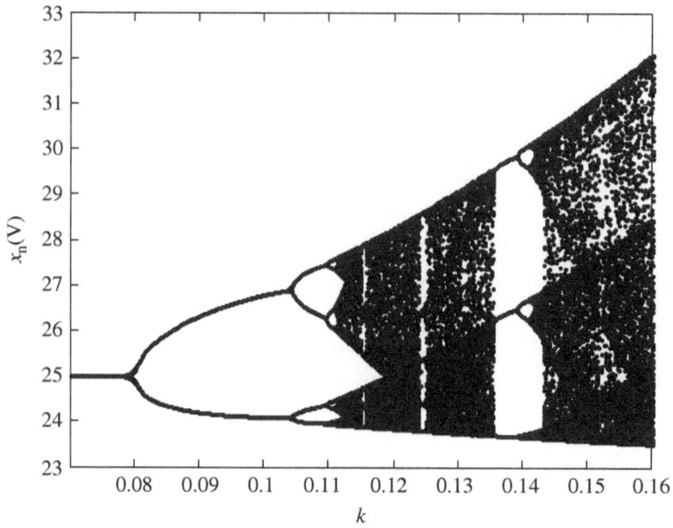

Figure 2.10 Bifurcation diagram of boost converter

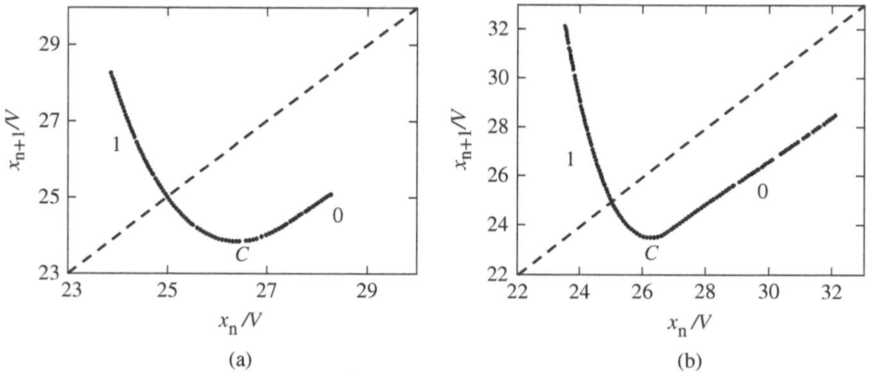

Figure 2.11 Return maps of boost converter at (a) $k = 0.12$ and (b) $k = 0.16$

A symbolic probability histogram with feedback gain $k = 0.16$ is shown in Figure 2.12, where $L = 8$ and the number of sampled data is 3000. We can find that the behavior of this converter is neither period nor random, and is a motion between them. Using Equation 2.10 further, the Shannon entropy scheme is shown in Figure 2.13. The period-doubling bifurcations can be observed clearly in Figure 2.13. Obviously, if the period is larger than 8, they won't be distinguished because the length L of a subsequence is 8. The following conclusions can be deduced from this figure.

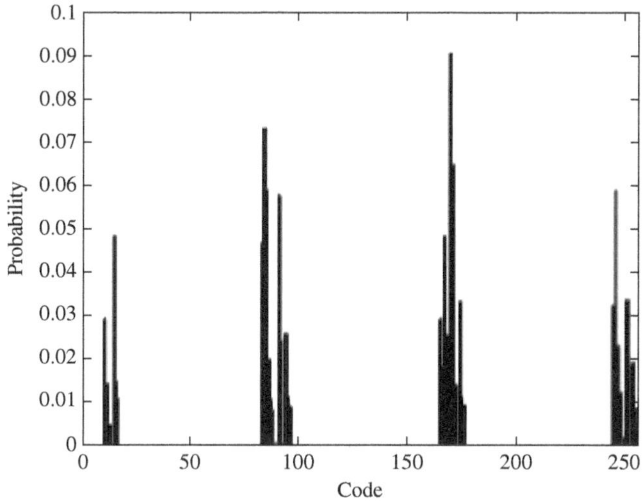

Figure 2.12 Symbolic probability histogram of DCM boost converter at $k = 0.16$

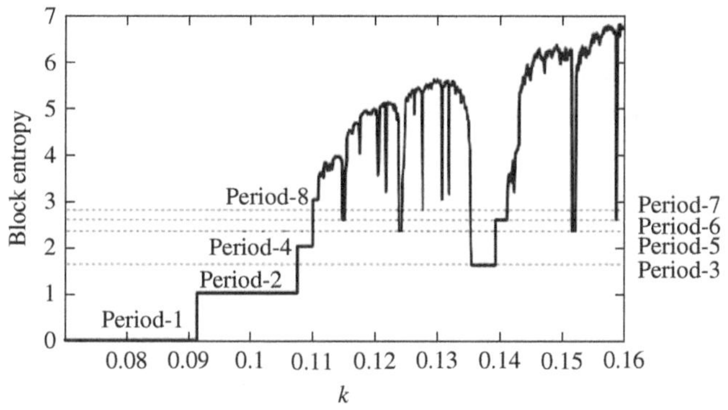

Figure 2.13 Shannon entropy versus the bifurcation parameter k. Here, the simulation step is 0.0001, subsequence length $L = 8$ and no. of discrete samples is 3000. The dashed lines are period-3, period-5, period-6, and period-7 respectively from the bottom up

1. When the entropy value keeps constant (horizontal line in Figure 2.13), the system is periodic.
2. The entropy value increases with the growth of chaos behavior.
3. The entropy value is less than L. These results strongly correlate with the properties of Shannon entropy introduced in Section 2.2.

The border between the periodicity and chaos of a symbolic series has been revealed in [24, 25]. The study results show that dynamics behavior approaches chaos

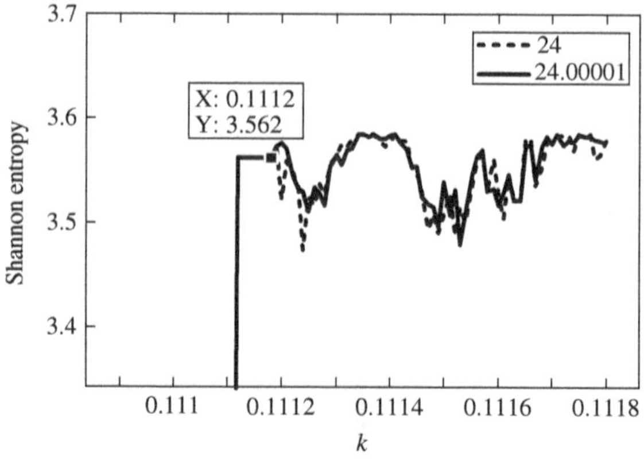

Figure 2.14 Shannon entropy diagram blown up with tiny different initial values (simulation step $k = 0.00001$)

via period-doubling at the *Feigenbaum point* when L reaches a much bigger value and $H^L \approx \log_2(L)/L$. This method has a limited application because the entropy per bit decays very slowly and the process requires a large computer memory. We propose here a new determination method for detecting the border between periodicity and chaos. Initial value sensitivity is an important property of chaos. For a symbolic series, a tiny difference of initial state values may cause significant disparity of Shannon entropy values. If the system is deterministic, we may confirm that it is chaotic because the random-like behavior of a deterministic system is chaotic. Two Shannon entropy curves of a boost converter with tiny different initial values are magnified in Figure 2.14. Two curves overlap completely when the system is periodic, then they stagger after $k > 0.1112$. Theoretically, for a DC–DC converter, an orbit will be extremely sensitive to tiny changes of the initial value when the system is in chaotic state. Moreover, this initial value sensitivity would be reflected by the value of entropy. Thus, chaos behavior appears when $k = 0.1112$, and it strongly correlates with the Lyapunov exponent of the boost converter (Figure 2.15). As we know, the Lyapunov exponent is a strictly mathematical method for detecting chaos motion, but it has two disadvantages:

1. It is difficult to obtain.
2. Long data series are required.

However, Shannon entropy is very simple because fewer samples are required and it can be quickly computed compared with the Lyapunov exponent. Therefore, our method can be easily applied to detect both bifurcation and chaos in a deterministic system.

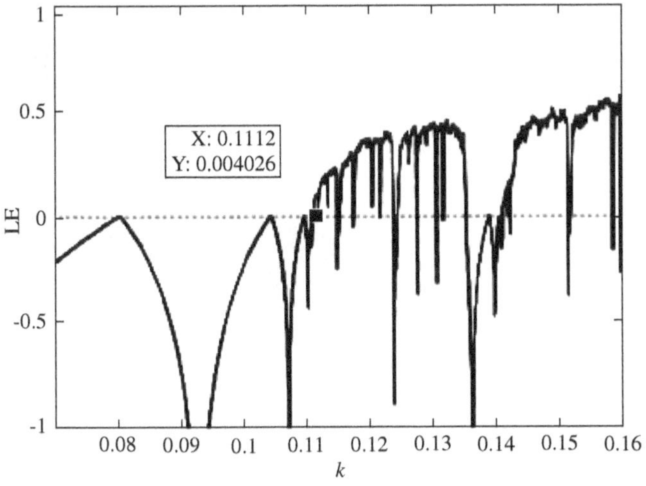

Figure 2.15 Lyapunov exponent of voltage feedback controlled boost converter

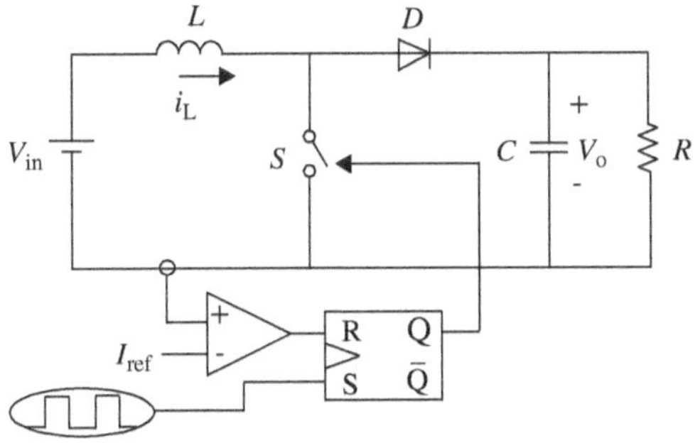

Figure 2.16 Boost converter with peak-current control

2.4.2 Border Collision Bifurcation and Chaos of DC–DC Converters

Peak-current control is the usual method applied in a boost converter. A boost converter with peak-current control is presented for analyzing nonlinearities. The schematic diagram is shown in Figure 2.16. The inductance current i_L is chosen as the variable parameter, which is compared with the reference current I_{ref}. The parameters are $E = 10$ V, $L = 1$ mH, $R = 40\,\Omega$, $C = 12\,\mu$F, $f = 10$ kHz, and the reference current I_{ref} changes from 1.0 to 5.5 A [26]. The operation of this converter has been studied by using an accurate iteration map, both period-doubling and border collision bifurcations can be observed in Figure 2.17. The border collision bifurcation is a common bifurcation often found in piecewise-smooth maps. We are aiming to analyze the border collision behavior by the method proposed earlier.

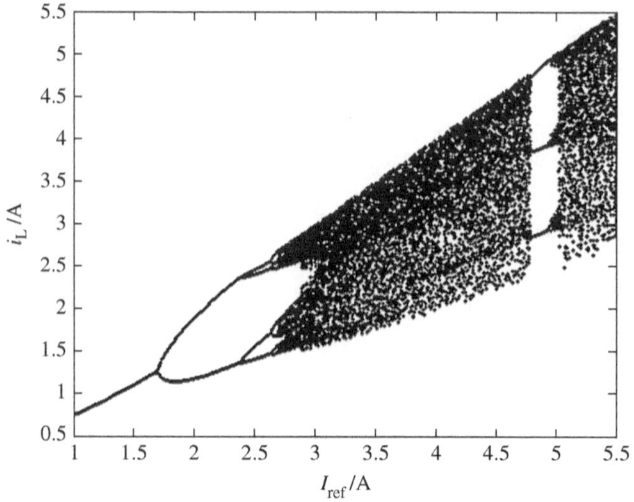

Figure 2.17 Bifurcation diagram of boost converter with peak current control

In [27], a borderline is given by the case where the inductance current i_L reaches the reference current I_{ref} at the arrival of the next clock pulse, and it is defined as

$$I_{border} = I_{ref} - \frac{E}{L}T \tag{2.11}$$

Let

$$i = I_{ref} - i_L \tag{2.12}$$

The borderline becomes

$$I_{border}' = I_{ref} - I_{border} = \frac{E}{L}T \tag{2.13}$$

The new border line I_{border}' is considered as a partition because it is constant independent of I_{ref}. The symbol s_n is defined as

$$s_n = \begin{cases} 1 & \text{if } i_n \leq 1 \\ 0 & \text{if } i_n > 1 \end{cases}$$

According to the iteration map, return maps of i are shown in Figure 2.18. These maps are similar to an inverted tent-like map and have a fixed critical point at $i_L = 1$ A.

Using Equation 2.10, the Shannon entropy scheme is shown in Figure 2.19. This result also strongly correlates with the largest Lyapunov exponent (Figure 2.20). Moreover, we notice that the entropy value jumps from 0 to 2 at $I_{ref} = 2.38$ A. This phenomenon indicates that this bifurcation isn't a standard bifurcation because the entropy value of standard bifurcation increases gradually. It further demonstrates that block entropy, whose partition is the borderline of a piecewise-smooth system, can be used to detect border collision bifurcation and chaos.

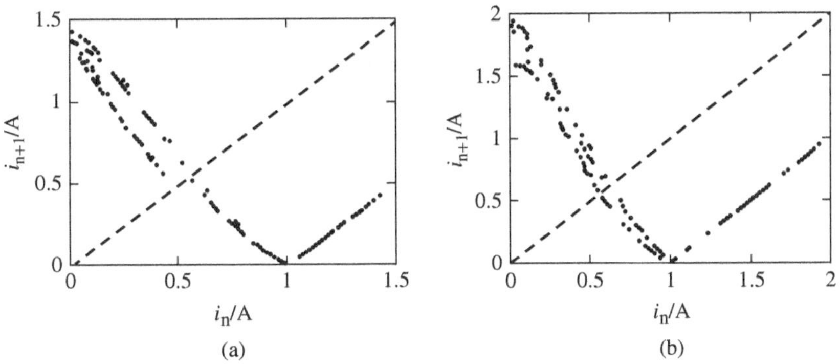

Figure 2.18 Return maps of i_n at (a) $I_{ref} = 1$ A and (b) $I_{ref} = 4$ A

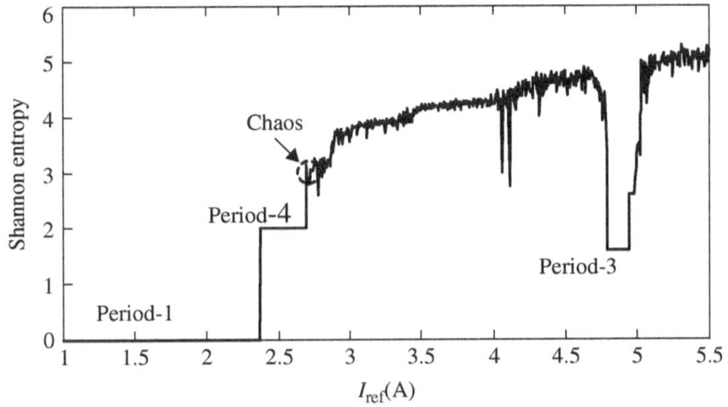

Figure 2.19 Shannon entropy versus the bifurcation parameter I_{ref} (simulation step is 0.0001, subsequence length $L = 6$, and no. of discrete samples is 3000)

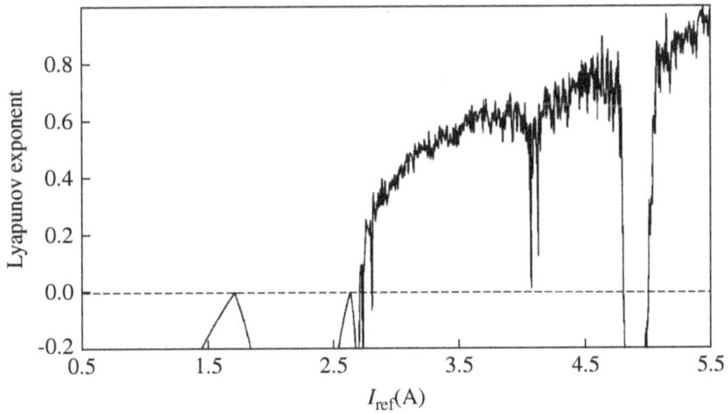

Figure 2.20 The largest Lyapunov exponent of a boost converter with peak-current control. Reprinted, with permission, from Deane, J.H.B, Chaos in a current-mode controlled dc-dc converter, IEEE Trans. Circuits Syst. I, and Aug./1992

2.5 Conclusions

Bifurcation and chaos are the main nonlinear phenomena in DC–DC switching converters, and the symbol method can analyze these complex behaviors. Thus, in this chapter, we introduce the symbolic method to analyze the nonlinear phenomena of DC–DC switching converters. First, an overview of the time series principle of discrete systems was shown. Then, we presented block entropy to analyze the bifurcation behaviors. Finally, we analyzed the bifurcation and chaos behavior of two boost converters via Shannon entropy. We find that initial value sensitivity is successfully used to distinguish chaos from periodic motions. The results show that this symbolic method strongly correlates with the Lyapunov exponent and can be calculated more easily and faster than the Lyapunov exponent.

References

[1] Dai, D. and Tse, C.K. (2004) Symbolic analysis of bifurcation in switching power converters: a novel method for detecting border collision. International Workshop on Nonlinear Circuit and Signal Processing (NCSP'04), Hawaii, pp. 5–7.

[2] Dai, D., Tse, C.K. and Ma, X. (2005) Symbolic analysis of switching systems: application to bifurcation analysis of DC-DC switching converters. *IEEE Trans. Circuits Syst. Regul. Pap.*, **52**(8), 1632–1643.

[3] Tse, C.K. and Dai, D. (2005) Symbolic analysis of bifurcation in switching power converters: a practical alternative viewpoint of border collision. *Int. J. Bifurcation Chaos*, **15**(7), 2263–2270.

[4] Maggio, G.M. and Galias, Z. (2002) Applications of symbolic dynamics to differential chaos shift keying. *IEEE Trans. Circuits Syst. Regul. Pap.*, **49**(12), 1729–1735.

[5] Khatkhate, A., Ray, A.E., Keller, S. *et al.* (2006) Symbolic time-series analysis for anomaly detection in mechanical systems. *IEEE Trans. Mechatron.*, **11**(4), 439–447.

[6] Godelle, J. and Letellier, C. (2000) Symbolic sequence statistics analysis for free liquid jet. *Phys. Rev. E*, **62**(6), 7973–7981.

[7] Kurths, J. (1995) Quantitative analysis of heart rate variability. *Chaos*, **5**(1), 88–94.

[8] Álvarez, G., Montoya, F., Romera, M. *et al.* (2003) Cryptanalysis of an ergodic chaotic cipher. *Phys. Lett. A.*, **311**(2), 172–179.

[9] Daw, C.S., Finney, C. and Tracy, E.R. (2003) A review of symbolic analysis of experimental data. *Rev. Sci. Instrum.*, **74**(2), 915–930.

[10] Wang, X.M., Zhang, B. and Qiu, D.Y. (2008) Symbolic time series characterization and block entropy analysis of DC-DC converters. *Acta Phys. Sinica*, **57**(10), 6112–6119.

[11] Wang, X.M., Zhang, B. and Qiu, D.Y. (2011) The quantitative characterization of symbolic series of a boost converter. *IEEE Trans. Power Electron.*, **26**(8), 2101–2105.

[12] Hao, B.L. (1989) *Elementary Symbolic Dynamics and Chaos in Dissipative Systems*, World Scientific, Singapore.

[13] Tang, X.Z., Tracy, E.R. and Brown, R. (1997) Symbol statistica and spatio-temporal systems. *Physica D*, **102**(3), 253–261.

[14] Tufillaro, N. (1999) Symbolic Dynamics in Mathematics, Physics, and Engineering, Hewlett-Packard Company, http://www.hpl.hp.com/techreports/1999/HPL-1999-28.html (accessed 7 June 2014).

[15] Bollt, E.M., Stanford, T., Lai, Y. *et al.* (2001) What symbolic dynamics do we get with a misplaced partition: on the validity of threshold crossings analysis of chaotic time-series. *Physica D*, **154**(3), 259–286.

[16] Letellier, C. (2008) Symbolic sequence analysis using approximated partition. *Chaos, Solitons, Fractals*, **36**(1), 32–41.

[17] Van Leeuwen, P., Cysarz, D., Lange, S. *et al.* (2007) Quantification of fetal heart rate regularity using symbolic dynamics. *Chaos*, **17**, 015119-1–015119-9.

[18] Tang, X.Z., Tracy, E.R., Boozer, A.D. *et al.* (1995) Symbol sequence statistics in noisy chaotic signal reconstruction. *Phys. Rev. E*, **51**(5), 3871–3889.

[19] Letellier, C., Meunier-Guttin-Cluzel, S. and Gouesbet, G. (2000) Topological invariants in period-doubling cascades. *J. Phys. A Math. Gen.*, **33**(9), 1809–1825.

[20] Wikipedia. Histogram, http://en.wikipedia.org/wiki/Histogram (accessed 7 June 2014)

[21] Wang, X.M. (2008) Symbolic time series characterization and block entropy analysis of DC-DC converters. *Acta Phys. Sinica*, **57**(10), 6112–6119.

[22] Wackerbauer, R., Witt, A. and Atmanspacher, H. (1994) A comparative classification of complexity measures. *Chaos, Solitons, Fractals*, **4**(1), 133–173.

[23] Tse, C.K. (1994) Flip bifurcation and chaos in three-state Boost switching regulator. *IEEE Trans. Circuits Syst. Reg. Pap.*, **41**(1), 16–23.

[24] Gramss, T. (1994) Entropy of the symbolic sequence for critical circle maps. *Phys. Rev. E*, **50**(4), 2616–2620.

[25] Grassberger, P. (1986) Toward a quantitative theory of self-generated complexity. *J. Theor. Phys.*, **25**(9), 907–938.

[26] Deane, J.B. (1992) Chaos in a current-mode controlled boost dc-dc converter. *IEEE Trans. Circuits Syst. I*, **39**(8), 680–683.

[27] Banerjee, S., Ranjan, P. and Grebogi, C. (2000) Bifurcations in two-dimensional piecewise smooth maps-theory and applications in switching circuits. *IEEE Trans. Circuits Syst. Regul. Pap.*, **47**(5), 633–643.

3

Complexity of the Nonlinear Behavior of DC–DC Converters

3.1 Introduction

Many nonlinear phenomena, such as bifurcations and chaos, have been identified in switching circuits and systems over the past 30 years [1–4]. When the system's parameters are varied, switching converters exhibit bifurcation phenomena and their topological structure will be changed cyclically. A series of methods has been used to analyze bifurcation behaviors in DC–DC converters, including using the maximal Lyapunov exponent to estimate the periodic orbits or chaotic orbits in the bifurcation process [5–7], analyzing the stability of the system based on the eigenvalue of the Jacobian matrix [8, 9], and detecting border collision bifurcation by estimating operation modes of the system [10–12]. On the other hand, bifurcation and chaos measurable properties, such as the power spectrum, Lyapunov exponents, fractal dimensions, and entropy, are intricate calculations and are too complex for engineering applications [13–15].

Recently, symbolic sequences have been used to analyze border collision and bifurcation phenomena in switching converters [16–21]. References [16–18] quantitatively analyze the nonlinear behaviors in switching converters using the twice-coarse-grained method. However, this method only adapts to the system of fewer operation modes for analyzing the nonlinear behaviors of the converters. Because the twice-coarse-grained method only exists as binary symbols "1" and "0", complex situations of more operation modes cannot be analyzed using this method. Tse and Dai in references [19–21] first presented the concepts of switching block and block switching sequence as a symbolic sequence method of switching converters to analyze the bifurcation phenomena across-the-board; this method is useful for distinguishing border collision bifurcation and standard bifurcations (period-doubling, Hopf, and saddle-node bifurcations). However, the waveforms of the system must

Chaos Analysis and Chaotic EMI Suppression of DC-DC Converters, First Edition. Bo Zhang and Xuemei Wang.

be observed to estimate standard bifurcations, and some bifurcation types – such as saddle node bifurcation, border collision bifurcations from period-n to period-n, period-m or chaos – have not been detected, so there are some limitations.

Complexity analysis in terms of the symbolic time sequence method gets the essence by explaining the inner structure of the system concretely. It is equal to the shortest sequence length of the figure structure or symbolic sequence of the system [22]. Yang in reference [16] adopted arithmetic complexity to analyze complex behaviors of the switching converter and described the working cycling and catastrophe point of the switching converters, which provided the theoretical basis for understanding the complexity characteristics of the converter. But reference [16] only provided the judging method for standard bifurcations, and border collision bifurcation was not distinguished.

Based on the limitations of the previous studies shown above, the concepts of duplicate symbolic sequence and weight complexity are put forward to analyze the bifurcation behaviors (including border collision bifurcation and standard bifurcation) of the switching converters in this chapter. According to the characteristics of DC–DC switching converters, the main switching block is proposed first in one switching period. Using the logical arithmetic method of computers that changes the decimal system from the binary system, the main symbolic sequence is established. Moreover, the secondary switching block of the converters and the secondary symbolic sequence are established using the coarse-grained symbolic sequence method. Thus, duplicate symbolic sequences and some criteria are constituted, which are utilized to analyze the border collision and bifurcation phenomena of the switching converters. Then the concept of weight complexity in the switching converters in terms of Lempel–Ziv (L–Z) complexity is put forward due to the main switching block incorporating the degree of complexity of each operation mode. Linking weight complexity and L–Z complexity, the complexity of the system and bifurcation behaviors are explained too. Finally, to show the effectiveness of the proposed concepts, we will illustrate the application procedure using the voltage-mode-controlled flyback converter as an example. But it should be noted that this method can be directly applied to other switching converters.

In this chapter, we will present the concepts of complexity and weight complexity to analyze bifurcation behaviors in DC–DC switching converters. First of all, we introduce L–Z complexity. We propose a new concept of weight complexity in terms of DC–DC switching converters; meanwhile, the method of complexity will be used to detect bifurcation behaviors in DC–DC converters. Next, the concepts of the main switching block and the main symbolic sequence in DC–DC switching converters is presented. The quantization values of the system's complexity in one switching period are obtained using the main switching block. Based on the main switching block and the coarse-grained symbolic sequence method, the secondary switching block and secondary symbolic sequence are established too. Therefore, the duplicate symbolic sequence is proposed to detect all of the main bifurcation types in DC–DC converters, and then the bifurcation types are detected by comparing the sequence before and after bifurcation.

3.2 Lempel–Ziv Complexity and Analysis of Nonlinear Behavior of DC–DC Converters Based on L–Z Complexity

3.2.1 Lempel–Ziv Complexity

As we all know, to get a more accurate result of the Lyapunov exponent and correlation dimension, tens of thousands of data should be calculated. However, this data acquisition is not easy in practical research areas such as life sciences, chemistry, or physics, because long data samples with low noise may be beyond the scope of the experiment. Hence, different kinds of definitions of different complexity are put forward by researchers in order to measure system dynamical features with fewer samples. Complexity is still a broad concept, and Lempel–Ziv complexity derived from symbolic dynamics is the most famous.

The demands on data to calculate the complexity is far less than for the Lyapunov exponent and correlation dimension, and often only a few hundreds to thousands of data can make a more robust estimate of systems with no sensitivity to noise. Lempel–Ziv complexity comes from the question of what is "complexity", proposed by Kolmogorov, Chaitin, and other researchers, which is the minimum amount of information represented by a sequence.

Kolmogorov once suggested that the algorithm complexity of a symbolic sequence should be determined by the length of the shortest computer program; but no one can give this algorithm. Although the shortest computer program cannot be obtained, Lempel and Ziv raised an algorithm that has an upper bound of the length of the computer program. Usually, it is called Lempel–Ziv complexity. Essentially, this algorithm measures the increase in speed of new modes as the length of the time sequence increases. In 1987, a detailed algorithm was given by Kasper and Schuster.

For a given finite symbol sequence $L = l_1\ l_2\ l_3 \ldots l_n$, Lempel and Ziv [23] have proposed a useful complexity measure C and have offered the related mathematical definitions and detailed derivations. The complexity measure can be estimated using the following algorithm [23, 24].

1. ε and θ are denoted as two subsequences of L and let $\varepsilon\theta$ be the concatenation of ε and θ, while sequence $\varepsilon\theta\pi$ is derived from $\varepsilon\theta$ after its last character is deleted (π denotes the operation of deleting the last character in the sequence). Let $V(\varepsilon\theta\pi)$ denote the vocabulary of all different subsequences of $\varepsilon\theta\pi$. At the beginning, $C = 1$, $\varepsilon = l_1$, and $\theta = l_2$, therefore, $\varepsilon\theta\pi = l_1$.
2. $\varepsilon = l_1\ l_2\ l_3 \ldots l_r$ and $\theta = l_{r+1}$, in general, so $\varepsilon\theta\pi = l_1\ l_2\ l_3 \ldots l_r$; if θ belongs to $V(\varepsilon\theta\pi)$, then θ is a subsequence of $\varepsilon\theta\pi$, not a new sequence.
3. Renew θ to be l_{r+1}, l_{r+2} and judge if θ belongs to $V(\varepsilon\theta\pi)$ or not.
4. Repeat the previous steps until θ does not belong to $V(\varepsilon\theta\pi)$. Now $\theta = l_1\ l_2\ l_3 \ldots l_{r+i}$ is not a subsequence of $V(\varepsilon\theta\pi) = l_1\ l_2\ l_3 \ldots l_{r+i-1}$, so we use the separator • to divide the sequence as $\varepsilon\theta = l_1\ l_2\ l_3 \ldots l_r \bullet l_{r+1}\ l_{r+2} \ldots l_{r+i}$ and increase C by one.
5. Thereafter, ε is renewed to be $\varepsilon = l_1\ l_2\ l_3 \ldots l_{r+i}$ and $\theta = l_{r+i+1}$.
6. The above procedure is repeated until θ is the last character. At this time the number of different subsequences in S, the measure of complexity, is C.

3.2.2 Analysis of Lempel–Ziv Complexity of Buck Converter

The buck converter under peak-current control is shown in Figure 3.1. The parameters used in the simulation are $R = 15\,\Omega$, $C = 50\,\mu F$, $L = 0.5\,mH$, $I_{ref} = 0.925\,A$. The simulation is based on an iterative map of inductor current for various values of input voltage E. Using the L–Z complexity algorithm, L–Z complexity is calculated via 1000 discrete points after ignoring the transient process, and the results are shown in Figures 3.2 and 3.3.

From the above converter, nonlinear dynamical behavior of discrete maps can be described by the L–Z complexity of a symbolic sequence with changing parameter [16]. Next, the bifurcation diagram of the inductor current (Figure 3.4) is shown to verify the accuracy of the nonlinear behavior of the switching converters by L–Z complexity. In Figure 3.4, let i_L and v_C at the nth switching moment be i_n and v_n respectively, and those at the next moment be i_{n+1} and v_{n+1} respectively.

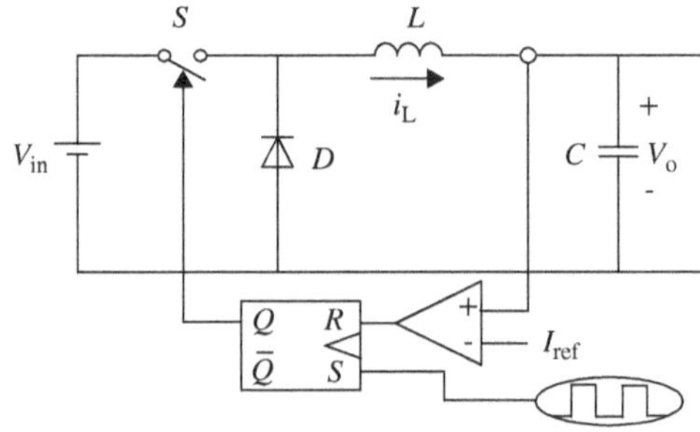

Figure 3.1 Buck converter with peak-current control

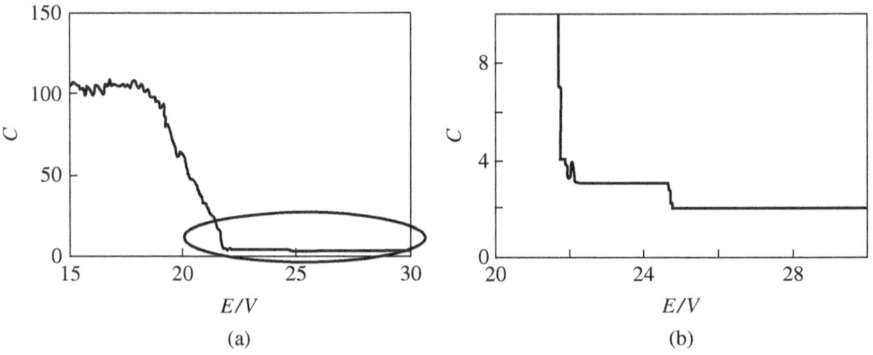

Figure 3.2 (a) L–Z complexity of a buck converter with E as the varied parameter and (b) the enlarged part of (a)

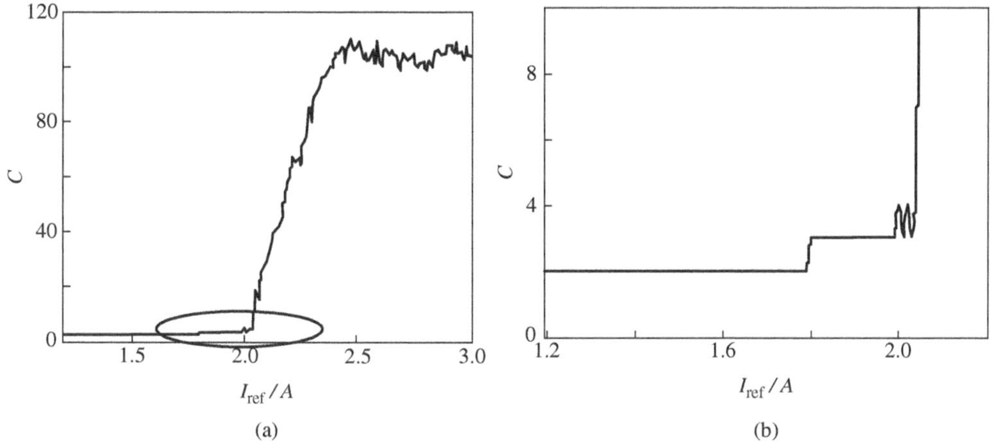

Figure 3.3 (a) L–Z complexity of a buck converter with I_{ref} as the varied parameter and (b) the enlarged part of (a)

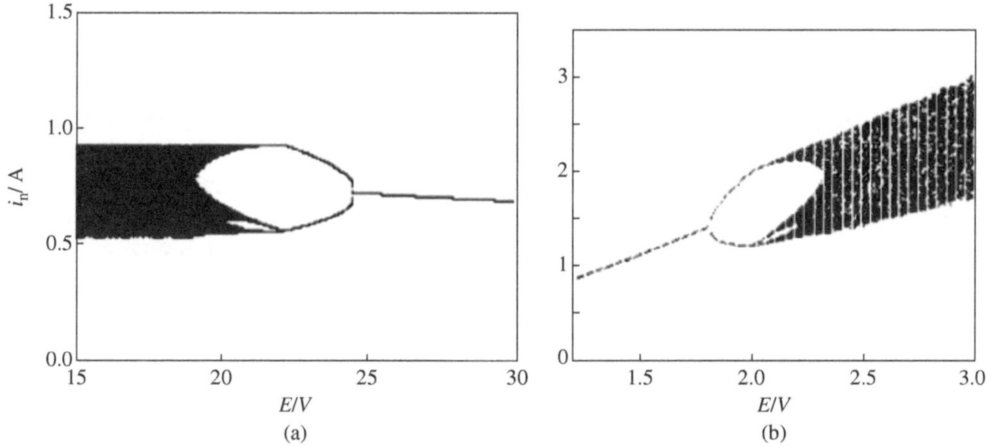

Figure 3.4 Bifurcation diagrams with (a) E and (b) I_{ref} as the bifurcation parameters

3.3 Switching Block of DC–DC Converters

The operation of a switching converter can be described as the orderly repetition of a fixed sequence of circuit topologies. There are 2^N possible sub-circuit topologies in DC–DC converters if the number of switching components (as switch and power diode) is N. Actually, not all sub-topologies will appear because the practical operation not only lies on the control method but also the conduction mode of the system. In general, a complete period of the system should have a process of absorbing and releasing energy. The function of some switching modes assists the system in absorbing energy through reactive (energy storing) components, such as inductors and capacitors, and one of the others assists the system in releasing energy. Especially, the switching mode of absorbing energy usually appears in one switching period first, and then the switching mode of releasing energy follows.

In traditional DC–DC converters, there are two switching elements (one switch and one diode). Thus the number of switching modes is equal to $2^2 = 4$, including $(S,D) = (\text{on,on})$, $(S,D) = (\text{on,off})$, $(S,D) = (\text{off,on})$, and $(S,D) = (\text{off,off})$, here S and D are switch and diode respectively, and on and off denote the switching elements turning on and turning off respectively. Different working modes of typical DC–DC converters are shown in Figures 3.5–3.8. In practice, however, switching mode $(S,D) = (\text{on,on})$ does not exist in the operational process of the converters. The system absorbs energy in switching mode $(S,D) = (\text{on,off})$, then it releases energy in switching modes $(S,D) = (\text{off,on})$ and $(S,D) = (\text{off,off})$. As said above, the absorbing switching mode $(S,D) = (\text{on,off})$ happens preferentially during one switching period, and then the releasing switching modes $(S,D) = (\text{off,on})$ and $(S,D) = (\text{off,off})$ occur. In addition, $(S,D) = (\text{off,off})$ happens after $(S,D) = (\text{off,on})$, because in the releasing state of the converters, D always turns on until the magnetic element releases all of its energy. Therefore, the switching modes sequence is always $(S,D) = (\text{on,off}) \rightarrow (S,D) = (\text{off,on}) \rightarrow (S,D) = (\text{off,off})$ during one switching period. It should be noted that the switching mode $(S,D) = (\text{off,off})$ does not occur in continuous conduction mode. Normally the system will not have more than three switching modes in one switching period. We use three binary coded digits B_1, B_2, and B_3 to express the combinations of three switching modes $(S,D) = (\text{on,off})$,

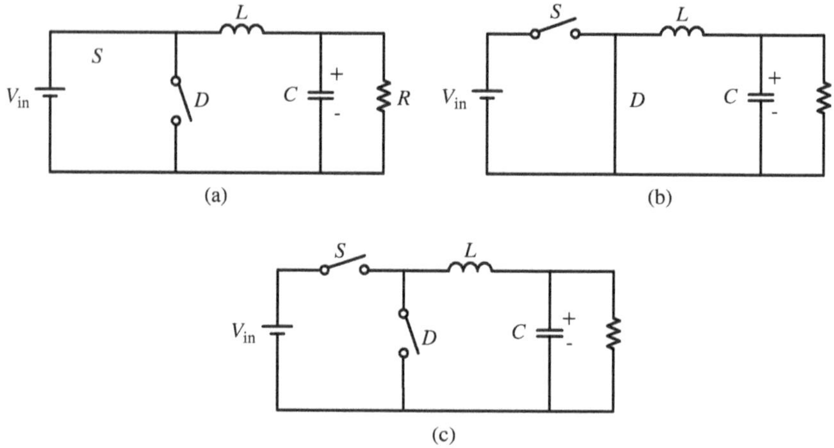

Figure 3.5 Buck converter at (a) $(S,D) = (\text{on,off})$, (b) $(S,D) = (\text{off,on})$, and (c) $(S,D) = (\text{off,off})$

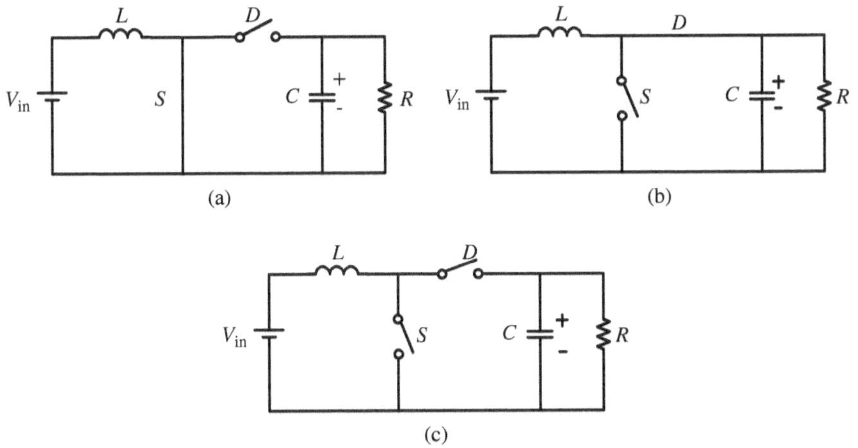

Figure 3.6 Boost converter at (a) $(S,D) = (\text{on,off})$, (b) $(S,D) = (\text{off,on})$, and (c) $(S,D) = (\text{off,off})$

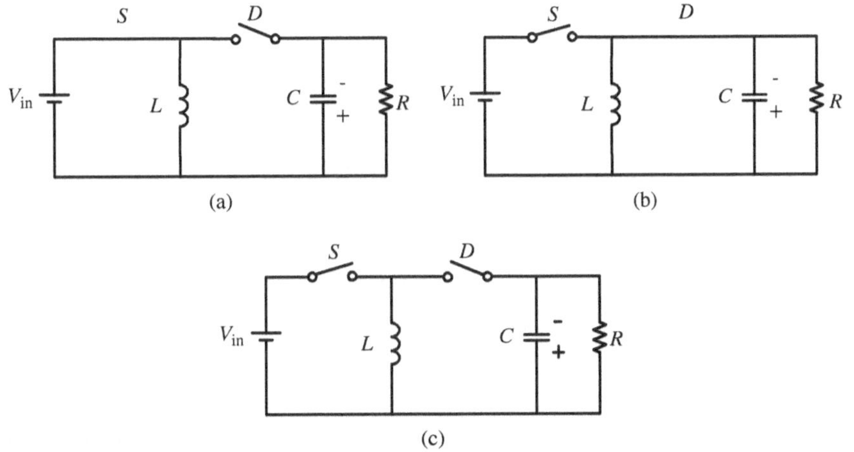

Figure 3.7 Buck–boost converter at (a) $(S,D) = (\text{on,off})$, (b) $(S,D) = (\text{off,on})$, and (c) $(S,D) = (\text{off,off})$

$(S,D) = (\text{off,on})$, and $(S,D) = (\text{off,off})$, respectively, in one switching period. Let $B_i = 0$ and $B_i = 1$ represent whether the responding switching mode appears or not during one switching period, respectively. Using the logical arithmetic method of the computer, we change the binary system to the decimal system and the main switching block is as follows

$$(B_1 B_2 B_3)_2 = (o)_{10} \tag{3.1}$$

here o is a decimal symbol and equates to 0, 1, 2, …, 7. For example, for the switching modes sequence $(S,D) = (\text{on,off}) \to (S,D) = (\text{off,on}) \to (S,D) = (\text{off,off})$, $B_1 = 1$,

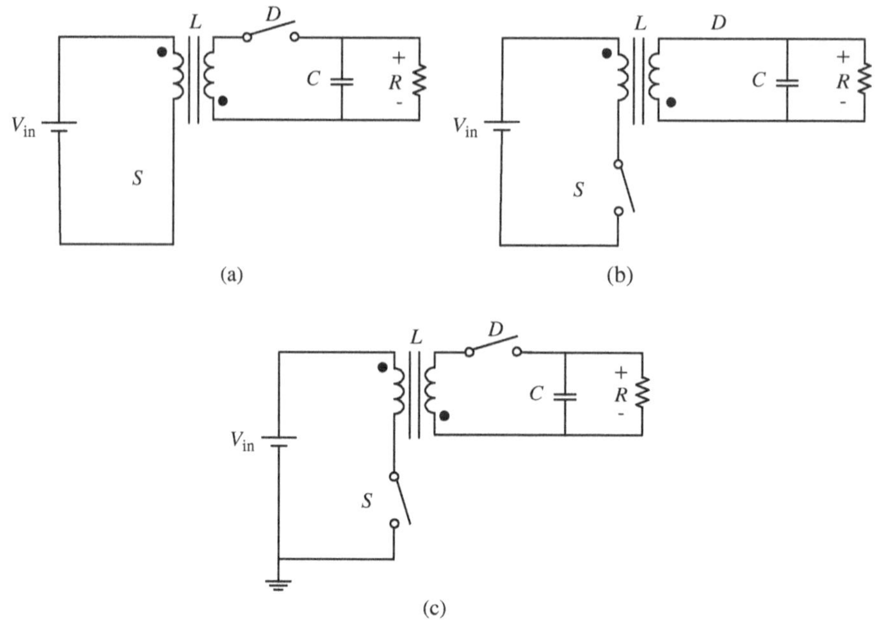

Figure 3.8 Flyback converter at (a) $(S,\ D) = (\text{on},\text{off})$, (b) $(S,D) = (\text{off},\text{on})$, and (c) $(S,D) = (\text{off},\text{off})$

$B_2 = 1$, and $B_3 = 1$, so the main switching block is $(111)_2 = 7$. For the switching modes sequence $(S,D) = (\text{on},\text{off}) \rightarrow (S,D) = (\text{off},\text{on}) \rightarrow \overline{(S,D) = (\text{off},\text{off})}$, $B_1 = 1$, $B_2 = 1$, and $B_3 = 0$, so the main switching block is $(110)_2 = 6$. For the switching modes sequence $(S,D) = (\text{on},\text{off}) \rightarrow \overline{(S,D) = (\text{off},\text{on})} \rightarrow \overline{(S,D) = (\text{off},\text{off})}$, $B_1 = 1$, $B_2 = 0$, and $B_3 = 0$, so the main switching block is $(100)_2 = 4$, and so on. Here we denote the non-appearing switching modes as $\overline{(S,D) = (\text{on\&off},\text{on\&off})}$. Obviously, the main switching block conceals the degree of the system's complexity in one switching period. During one switching period, for instance, $o = 1$ expresses that one switching mode takes place and $o = 7$ expresses that three switching modes and two switching behaviors take place.

3.4 Weight Lempel–Ziv Complexity and Analysis of Nonlinear Behavior of DC–DC Converters Based on Weight L–Z Complexity

L–Z complexity is a measure of the complexity of any given finite-length sequence of symbols, and it has been used in various fields for its merit of convenience of calculation for a short sequence study. Unfortunately, L–Z complexity cannot be used to fully explain the complexity of piecewise-smooth systems. For this reason, the conception of weight Lempel–Ziv complexity is put forward [25, 26].

3.4.1 Weight Lempel–Ziv Complexity

As mentioned in Section 3.3, the switching block conceals the degree of the system's complexity in one switching period. On the other hand, L–Z complexity only counts the quantity of subsequence, so it cannot distinguish the situation that the system exhibits border collision bifurcation from period-n to period-n. Therefore, for the main symbolic sequence O, we delete the first subsequence or subsequences of O and let the new main symbolic sequence have the descending order. Based on L–Z complexity, the sum of the maximal main switching blocks of each subsequence in the new main symbolic sequence is denoted as the weight complexity C_W. We present the following example about the main symbolic sequence $O=(676767676\ldots)_\infty=(67)_\infty$ to explain this concept.

1. First, $O=(676767676\ldots)$ is renewed as $O=(76767676\ \ldots)$.
2. Second, let $\varepsilon=o_1=7$, $\theta=o_2=6$, $\varepsilon\theta=76$, $\varepsilon\theta\pi=7$, so θ is not the subsequence of $\varepsilon\theta\pi$, and then adopt the separator • on $\varepsilon\theta$, so $\varepsilon\theta=7\bullet6\bullet$.
3. Moreover, let $\varepsilon=o_1o_2=76$, $\theta=o_3=6$, $\varepsilon\theta=767$, $\varepsilon\theta\pi=76$, so θ is the subsequence of $\varepsilon\theta\pi$ and $\varepsilon\theta=7\bullet6\bullet7$.
4. And then let $\varepsilon=o_1o_2=76$, $\theta=o_3o_4=76$, $\varepsilon\theta=7676$, $\varepsilon\theta\pi=767$, so θ is the subsequence of $\varepsilon\theta\pi$ and $\varepsilon\theta=76\bullet76$.
5. Steps (2) and (3) are iterated, and then we obtain $O=7\bullet6\bullet76767676\ \ldots$
6. So this main symbolic sequence has three subsequences. The maximal blocks in three subsequences are 7, 6, and 7, respectively. Thus the weight complexity is $C_W=7+6+7=20$.

3.4.2 Weight Lempel–Ziv Complexity of Buck Converter

Weight Lempel–Ziv complexity is used to analyze the current feedback controlled buck converter (Figure 3.1). The circuit parameters are $E=20\,\text{V}$, $L=3.3\,\text{mH}$, $C=1000\,\mu\text{F}$, $R=19\,\Omega$, $T=400\,\mu\text{s}$, and I_{ref} is the bifurcation parameter. Based on the algorithm of the weight Lempel–Ziv complexity introduced above, the C_W of the buck converter is shown in Figure 3.9.

As can be seen from Figure 3.9, the weight Lempel–Ziv complexity equals $C_w=7+7=14$ as $I_{ref}\in(0.15\,\text{A},\ 0.2779\,\text{A})$. Obviously, the switching mode is 7 and the system is in period-1 orbit when I_{ref} is in this range. As $I_{ref}=0.2779\,\text{A}$, C_w decreases from 14 to 12, and this illustrates that the border collision phenomena take place beside the direct transition from switching mode 7 to switching mode 6. Switching mode 6 is kept unchanged when $I_{ref}\in(0.2779\,\text{A},\ 0.9007\,\text{A})$. Next, the weight Lempel–Ziv complexity rises from $C_w=6+6=12$ to $C_w=7+6+7=20$, and this indicates that the converter is in period-2 orbit composed together with switching mode 7 and 6. As I_{ref} increases, C_w falls to 18 at $I_{ref}=1.1578\,\text{A}$, and this also indicates that the converter is in period-2 orbit composed together with switching mode 7 and 4. With continuously increasing I_{ref}, $S=7\bullet4\bullet6\bullet47\bullet46474\ \ldots$ and $C_W=7+4+6+7+7=31$ in $I_{ref}\in(1.1947\,\text{A},\ 1.2625\,\text{A})$, this illustrates that

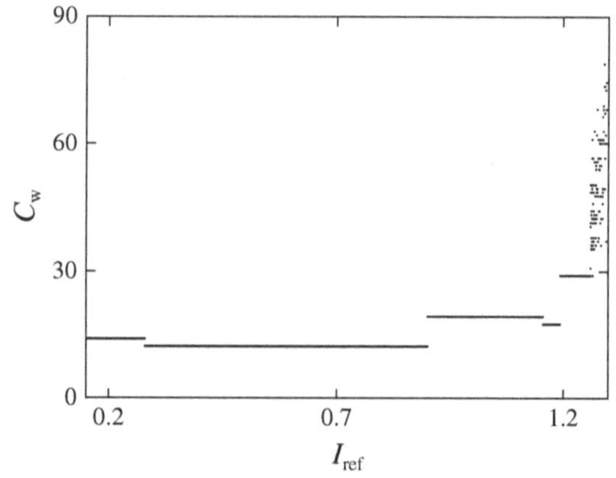

Figure 3.9 Weight Lempel–Ziv complexity of a buck converter with peak-current control

the system is in period-4 orbit with the switching modes sequence $7 \rightarrow 4 \rightarrow 6 \rightarrow 4$. When $I_{ref} > 1.2625$ A, the weight Lempel–Ziv complexity gets larger and more irregular, and it can be seen that the system has entered a chaotic state. Additionally, the inductance currents as shown in Figure 3.10 are explained as well. Therefore, the weight L–Z complexity can demonstrate the nonlinear behavior and operation complexity of a system.

3.4.3 Qualitative Analysis of Bifurcation Phenomena Based on Complexity

The main switching block contains the information of complexity, so we use weight complexity to analyze the main symbolic sequence. On the other hand, the secondary switching block is defined from the voltage level and only as a symbol, so we use L–Z complexity to analyze the secondary symbolic sequence. Let the weight complexity of the main symbolic sequence be C_W and let the Lemper–Ziv complexity of the secondary symbolic sequence be C in DC–DC switching converters. The types of bifurcation are distinguished via the following criteria.

1. When C changes but C_W does not change, the system exhibits standard bifurcations. If C increases but the increment of C is small, the system will exhibit period-doubling bifurcation. If C decreases but the decrement of C is more, the system will exhibit saddle node bifurcation. If C increases and the increment of C is more, the systems will exhibit Hopf bifurcation.
2. When C does not change but C_W changes, the system exhibits border collision bifurcation and jumps from one periodic orbit to the other periodic orbit that has the same periodic number as the former.
3. When C and C_W both change, the system exhibits border collision bifurcation and jumps from a periodic orbit to another periodic orbit that has a different periodic number of the former or chaotic orbit.

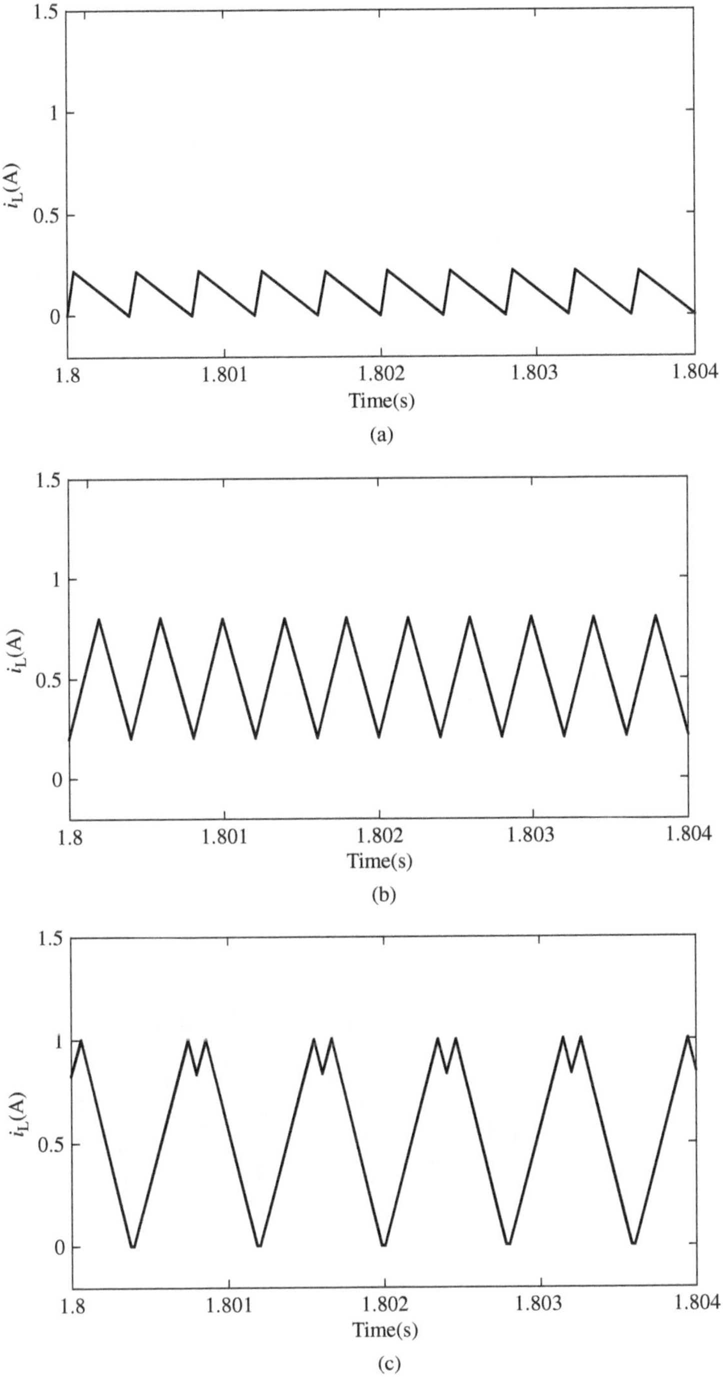

Figure 3.10 Inductor current waveforms at (a) $I_{ref} = 0.2$ A, (b) $I_{ref} = 0.8$ A, (c) $I_{ref} = 1$ A, (d) $I_{ref} = 1.16$ A, (e) $I_{ref} = 1.2$ A, and (f) $I_{ref} = 1.3$ A

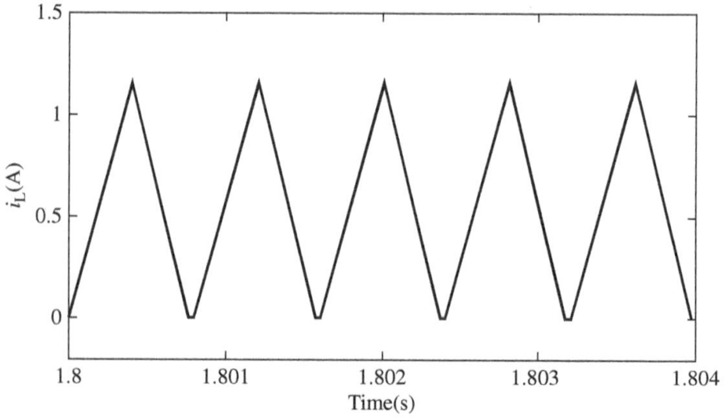

(d)

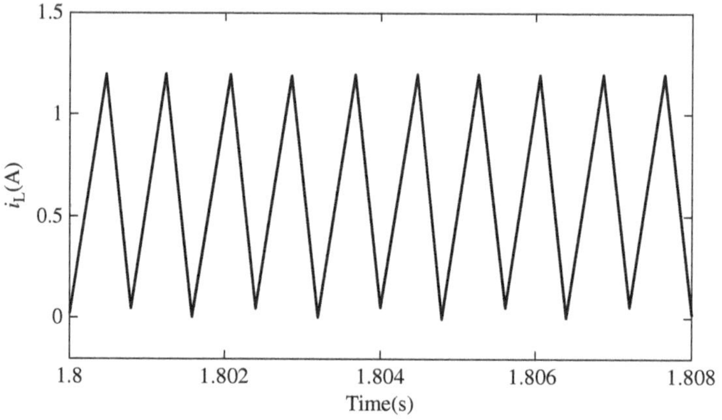

(e)

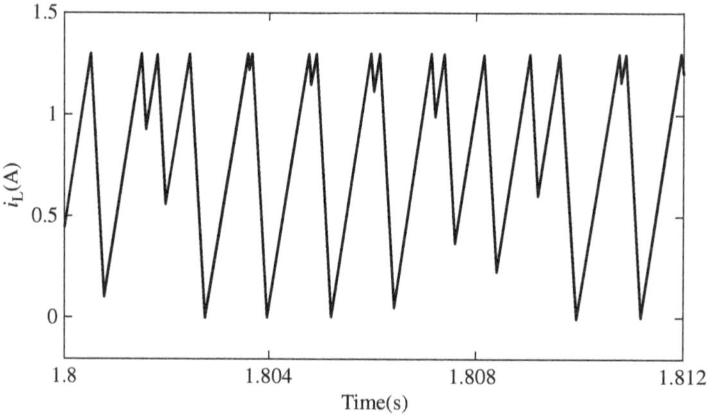

(f)

Figure 3.10 (*continued*)

3.5 Duplicate Symbolic Sequence and Complexity

3.5.1 Main Switching Block and Main Symbolic Sequence

According to the characteristics of DC–DC switching converters, the concept of the novel main switching block is presented [26].

Let the state variables of the system be $X = (x_1, x_2)^T$, here x_1 is a variable about current and x_2 is a variable about voltage. Through stroboscopic sampling, by observing the state variables at the beginning of each control period, when the system enters into stationary state we sample and obtain a time sequence of two dimensions

$$X_0, X_1, X_2, \cdots, X_n, \cdots \tag{3.2}$$

According to references [10, 11], DC–DC switching converters are piecewise-smooth systems and their state-spaces can be divided into several regions. Moreover, the switching behaviors of the system in each state region are decided during one switching period, that is to say the main switching block is clear in the decided region. Thus, Equation 3.2 can be changed to the main symbolic sequence

$$O = (o_0 o_1 o_2 \cdots o_n \cdots) \tag{3.3}$$

Here, o_i expresses the main switching block of X_i. According to the definition of references [19–21], $(o_0 o_1 o_2 \ldots o_m)_n$ is denoted as a finite main symbolic sequence which repeats the main symbolic sequence $(o_0 o_1 o_2 \ldots o_m)$ n times. Additionally, a periodic main symbolic sequence is denoted as $(o_0 o_1 o_2 \ldots o_m)_\infty$, and an aperiodic main symbolic sequence is denoted as (∞).

3.5.2 Secondary Switching Block and Secondary Symbolic Sequence

As shown before, main symbolic sequence is based on the main switching block, and the main switching block only considers how the switching behaviors change and what switching modes occur in one switching period. Therefore, some important information from the original dynamics system used as the main symbolic sequence in switching converters is missing. Hence, we present the concepts of the secondary switching block and secondary symbolic sequence based on the main symbolic sequence as follows.

Here we use the coarse-grained symbolic sequence method to establish the secondary switching block. For time sequence $X_i (i = 0, 1, 2, \ldots, n)$ of Equation 3.2, we let V_{max} and V_{min} be the maximal and minimal values of each x_{2i} (x_{2i} is the second variable of X_i) respectively. And then we let

$$Q = (\mathrm{V}_{max} - \mathrm{V}_{min})/(P + 1) \tag{3.4}$$

here P + 1 is the coarse-grained degree. From Equation 3.4, we can get the division point

$$a(j) = \mathrm{V}_{max} - Q_j (j = 1, 2, \ldots, P) \tag{3.5}$$

So, data points V_{max}, $a_{(1)} \ldots a_{(j)}$ and V_{min} make the time sequence X_i divided into P layers: $V_{max} \rightarrow \underline{a}(1)$ is the first layer, $a(1) \rightarrow a(2)$ is the second layer, …, $a(P-1) \rightarrow V_{min}$ is the *Ps* layer. If x_{2i} is at *ks* layer, the secondary switching block of X_i will be

$$l_i = k - 1 \; (k = 1, 2, \; \ldots \; , P) \tag{3.6}$$

and the corresponding secondary symbolic sequence of the system will be

$$L = (l_0 l_1 l_2 \; \ldots \; l_n \; \ldots) \tag{3.7}$$

Figure 3.11 is a thrice coarse-grained example to explain the coarse-grained method. According to the characteristics of DC–DC switching converters, if the main symbolic sequence of the system is periodic, we will let P of the Equation 3.4 be equal to the factorial of periodic number. Whereas if the main symbolic sequence of the system is aperiodic, we will let $P = 1$. For example, $P = m!$ for $O = (o_0 o_1 o_2 \ldots o_{m-1})_\infty$ and $P = 1$ for $O = (\infty)$.

3.5.3 Duplicate Symbolic Sequence

In summary, we obtain the main symbolic sequence and the secondary symbolic sequence of the DC–DC switching converters as shown in Equations 3.3 and 3.7 respectively. Let

$$s_i = o_i^{\,li} \tag{3.8}$$

be the switching block of the system and

$$S = (s_0 s_1 s_2 \; \ldots \; s_n \; \ldots) \tag{3.9}$$

be the duplicate symbolic sequence of the system. Similarly, we denote by $(s_0 s_1 s_2 \ldots s_m)_n$ a finite duplicate symbolic sequence which repeats the duplicate symbolic sequence $(s_0 s_1 s_2 \ldots s_m)$ n times. Moreover, a periodic duplicate symbolic sequence is denoted as $(s_0 s_1 s_2 \ldots s_m)_\infty$, and an aperiodic duplicate symbolic sequence as (∞).

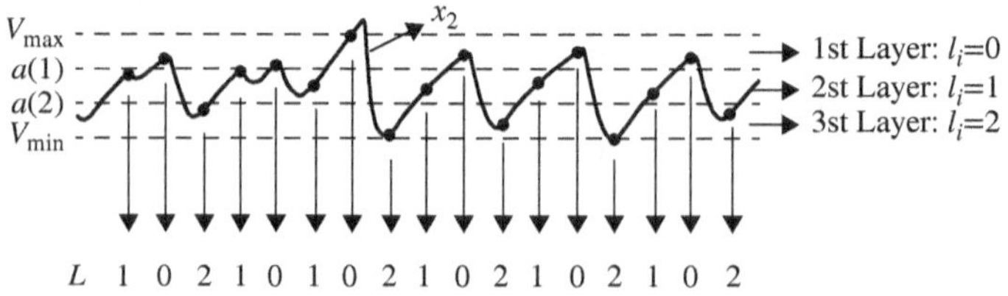

Figure 3.11 Second switching block and symbolic second sequence using thrice coarse-grained method

3.5.4 *Analysis of Border Collision and Bifurcation in DC–DC Converters Based on Duplicate Symbolic Sequence*

According to the definition of references [19–21], we consider a switching converter, which is described by the iterative map

$$X_n = f(X_{n-1}, \mu) \tag{3.10}$$

where μ is a parameter. When the converter undergoes a standard bifurcation as μ is varied, the form of f remains unchanged before and after the bifurcation. Whereas when the converter undergoes a border collision as μ is varied, the form of f alters at the bifurcation [21]. In our duplicate symbolic sequence, the main switching block is based on the form of f. In other words, a change of the form of f actually means a change of the main switching block.

As mentioned above, the secondary symbolic sequence is defined based on the main symbolic sequence, so we must analyze the main symbolic sequence first and then analyze the secondary symbolic sequence. On the other hand, the main symbolic sequence and secondary symbolic sequence constitute duplicate symbolic sequences; hence the sequence that we use in this symbolic method to analyze the bifurcation behaviors in DC–DC switching converters is $S \rightarrow O \rightarrow L$. Here we denote P_S, P_O, and P_L as the periodic number of periodic sequences S, O, and L, respectively. For instance, $P_S = m+1$ for $S = (s_0 s_1 s_2 \ldots s_m)_\infty$ and $P_S = +\infty$ for $S = \infty$.

Let the bifurcation parameter be $\mu (\mu \in R)$ in DC–DC switching converters. Suppose $\mu_1 < \mu_c < \mu_2$, where μ_c is the critical value. Denote the duplicate, main, secondary symbolic sequence, the periodic number of duplicate, main, and secondary symbolic sequence, respectively, by S_i, O_i, L_i, P_{Si}, P_{oi}, and P_{Li} for $\mu = \mu_i$ with $i = 1$ or 2. Once a bifurcation occurs, the relationship between S_1 and S_2 will be $S_1 \neq S_2$. But the type of bifurcation must relate to O, L, and their periodic numbers.

3.5.4.1 Detecting Standard Bifurcations in DC–DC Switching Converters Based on Duplicate Symbolic Sequences

As mentioned above, when the system undergoes standard bifurcations, the form of f remains unchanged before and after the bifurcation. Thus the main symbolic sequence of all the standard bifurcation is not changed, namely $O_1 = O_2$ and $P_{O1} = P_{O2}$. On the other hand, if O does not change but L changes, the system will lose its stability at a periodic orbit and will enter the other orbit. So we can detect standard bifurcation based on the change of L.

1. *Detecting period-doubling bifurcation.* When the system exhibits period-doubling bifurcation, the secondary symbolic sequence of the system will change. Figure 3.12 shows an example of period-doubling bifurcation from period-1 of $O = (7)_\infty$ to period-2 of $O = (7)_\infty$. As shown in Figure 3.12, the main switching block and main symbolic sequence are same, namely $P_{O1} = P_{O2} = 1$, so $P = 1$ and the coarse-grained degree in Equation 3.4 is $P + 1 = 2$. But there is only one

layer in the period-1 case because of $V_{max} = V_{min}$, whereas there are two layers in the period-2 case because of $V_{max} \neq V_{min}$. Hence there exists only one secondary switching block to detect period-1 and two secondary switching blocks are used to detect period-2. And then $L_1 = (0)_\infty$ and $P_{L1} = 1$ for period-1 and $L_2 = (01)_\infty$ and $P_{L1} = 2$ for period-2 as shown in Figure 3.2. In summary, if the system exhibits period-doubling bifurcation, there will be $S_1 \neq S_2$, $2P_{S1} = P_{S2}$, $O_1 = O_2$, $P_{O1} = P_{O2}$, $L_1 = L_2$, and $2P_{L1} = P_{L2}$.

2. *Detecting saddle node bifurcation.* Saddle node bifurcation is also termed tangent bifurcation, which occurs in smooth systems, so the main switching block in the bifurcation process remains unchanged, namely $O_1 = O_2$ and $P_{O1} = P_{O2}$. Before the saddle node bifurcation, the system works in intermittent chaos, consequently the secondary symbolic sequence is an infinite sequence and its periodic number is 1. After the saddle node bifurcation, the system works in a periodic state, consequently the secondary symbolic sequence is a periodic sequence and its periodic number is a fixed quantity. In summary, if the system exhibits a saddle node bifurcation, there will be $S_1 \neq S_2$, $P_{S1} \gg P_{S2}$, $O_1 = O_2$, $P_{O1} = P_{O2}$, $L_1 \neq L_2$, and $P_{L1} \gg P_{L2}$.
3. *Detecting Hopf bifurcation.* If the system exhibits Hopf bifurcation, there will be $S_1 \neq S_2$, $P_{S1} \ll P_{S2}$, $O_1 = O_2$, $P_{O1} = P_{O2}$, $L_1 \neq L_2$, and $P_{L1} \ll P_{L2}$. After the Hopf bifurcation, the system works in a quasi-periodicity state, consequently the secondary symbolic sequence is an infinite sequence and its periodic number is $+\infty$ as shown in Figure 3.13.

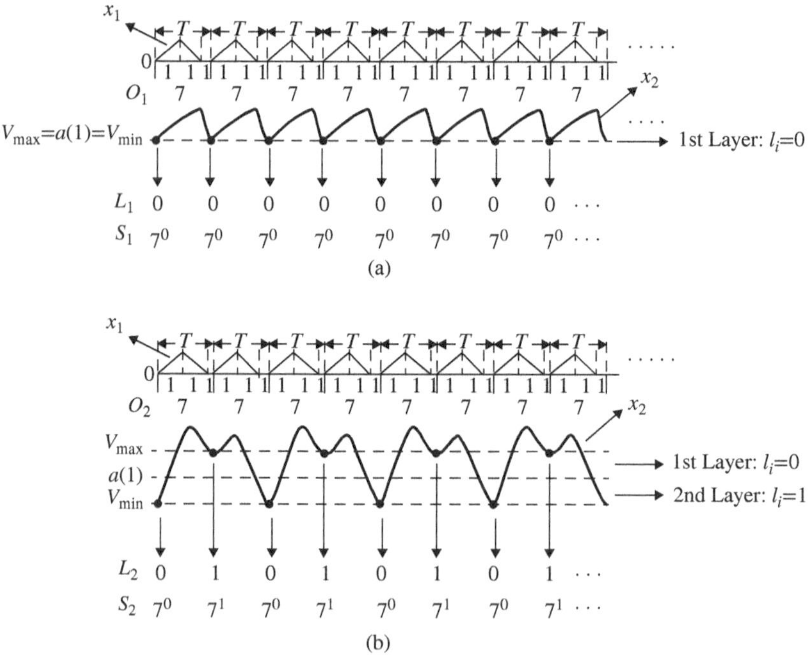

Figure 3.12 Analysis of period-doubling bifurcation using duplicate symbolic sequence for (a) period-1 of $o = 7$ and (b) period-2 of $o = 7$

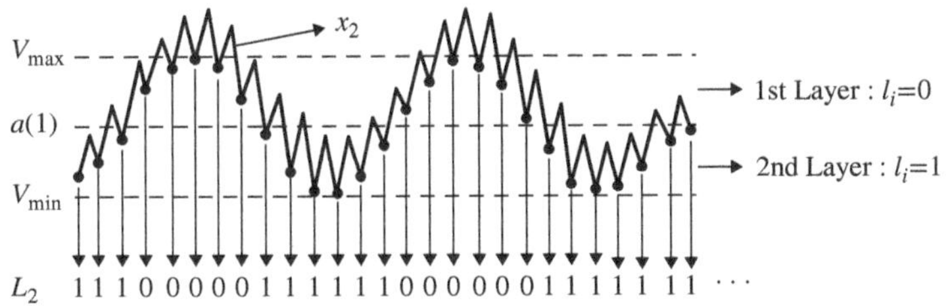

Figure 3.13 Analysis of quasi-periodicity in the duplicate symbolic sequence

3.5.4.2 Detecting Border Collision Bifurcations in DC–DC Converters Based on Duplicate Symbolic Sequences

As mentioned above, when the system undergoes border collision bifurcation, the form of *f* changes before and after the bifurcation. Thus the main symbolic sequence of border collision bifurcation is changed, namely $O_1 \neq O_2$, but the relationship between P_{O1} and P_{O2} may be equal or not. The secondary symbolic sequence is established based on the periodic number of the main symbolic sequence, so L varies with the corresponding O. According to the characteristics of border collision bifurcation, we summarize the following cases of border collision bifurcation.

1. *Period-n to period-n*: If $S_1 \neq S_2$, $O_1 \neq O_2$, $L_1 \neq \& = L_2$, $P_{S1} = P_{S2}$, $P_{O1} = P_{O2}$, and $P_{L1} = P_{L2}$, border collision will make the system jump from a period-n orbit to the other period-n orbit.
2. *Period-n to period-m*: If $S_1 \neq S_2$, $O_1 \neq O_2$, $L_1 \neq L_2$, $P_{S1} \neq P_{S2}$, $P_{O1} \neq P_{O2}$, and $P_{L1} \neq P_{L2}$ (here all of the periodic numbers are fixed quantities), border collision will make the system jump from a period-n orbit to another periodic orbit but the periodic number of this orbit is not equal to *n*.
3. *Period-n to chaos*: If $S_1 \neq S_2$, $O_1 \neq O_2$, $L_1 \neq L_2$, $P_{S1} \ll P_{S2}$, $P_{O1} \ll P_{O2}$, and $P_{L1} \ll P_{L2}$ (here P_{S1}, P_{O1}, and P_{L1} is the fixed quantities but $P_{S2} = P_{O2} = P_{L2} = +\infty$), border collision will make the system jump from a period-n orbit to chaos.

In summary, all of the bifurcation cases can be detected using the duplicate symbolic sequence as shown in Table 3.1.

3.6 Applied Example

In this section, the duplicate symbolic sequence and complexity are applied to analyze the bifurcation behaviors in a voltage-mode-controlled flyback converter as shown in Figure 3.14. The details of traditional methods for this converter were investigated in our previous study [11]. The operation of the flyback converter under study can be briefly described as follows. The switching of the MOSFET is controlled by the pulse width modulated (PWM) voltage loop. It is achieved by comparing a control voltage

Table 3.1 Detecting types of bifurcation based on duplicate symbolic sequences

Types of bifurcations		$S_1\&S_2$	$P_{S1}\&P_{S2}$	$O_1\&O_2$	$P_{O1}\&P_{O2}$	$L_1\&L_2$	$P_{L1}\&P_{L2}$
Standard bifurcation	Period-doubling	$S_1 \neq S_2$	$2P_{S1} = P_{S2}$	$O_1 = O_2$	$P_{O1} = P_{O2}$	$L_1 \neq L_2$	$2P_{L1} = P_{L2}$
	Saddle node	$S_1 \neq S_2$	$P_{S1} \gg P_{S2}$	$O_1 = O_2$	$P_{O1} = P_{O2}$	$L_1 \neq L_2$	$P_{L1} \gg P_{L2}$
	Hopf	$S_1 \neq S_2$	$P_{S1} \ll P_{S2}$	$O_1 = O_2$	$P_{O1} = P_{O2}$	$L_1 \neq L_2$	$P_{L1} \ll P_{L2}$
Border collision bifurcation	Period-n to period-n	$S_1 \neq S_2$	$P_{S1} = P_{S2}$	$O_1 \neq O_2$	$P_{O1} = P_{O2}$	$L_1 \neq \& = L_2$	$P_{L1} = P_{L2}$
	Period-n to period-m	$S_1 \neq S_2$	$P_{S1} \neq P_{S2}$	$O_1 \neq O_2$	$P_{O1} \neq P_{O2}$	$L_1 \neq L_2$	$P_{L1} \neq P_{L2}$
	Period-n to chaos	$S_1 \neq S_2$	$P_{S1} \ll P_{S2}$	$O_1 \neq O_2$	$P_{O1} \ll P_{O2}$	$L_1 \neq L_2$	$P_{L1} \ll P_{L2}$

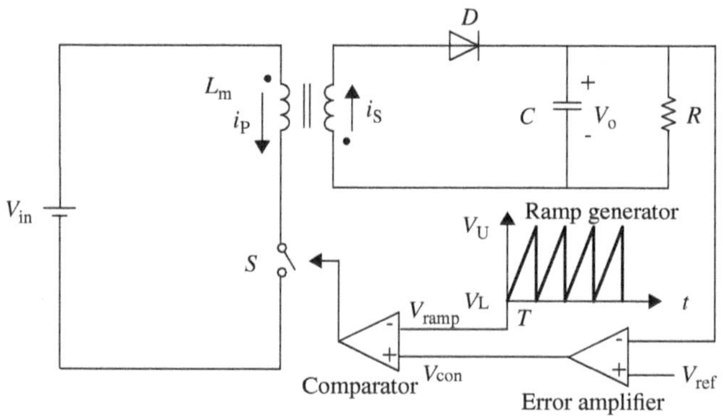

Figure 3.14 Voltage-mode-controlled flyback converter

V_{con} with an independent sawtooth ramp voltage V_{ramp}. The expression for the control voltage is

$$V_{con}(t) = k(V_{ref} - v(t)) \tag{3.11}$$

where k is the gain of the error amplifier, V_{ref} is the reference voltage, and $v(t)$ is the instant value of the capacitor voltage. The sawtooth ramp voltage is illustrated in the form

$$V_{ramp}(t) = V_L + \frac{V_U - V_L}{T}t \quad t \in [0, T) \tag{3.12}$$

where T is the control period, V_U and V_L are the upper and lower threshold voltages respectively. S is turned on when $V_{con}(t) > V_{ramp}$, and is turned off when $V_{con}(t) < V_{ramp}$. There is not a universal variable like the inductor current in a buck converter, so we use total ampere-turns ρ of the primary and secondary windings as one of the state variables to measure the situation of the transformer. The expression of ρ is

$$\rho(t) = N_p i_p(t) + N_s i_s(t) \tag{3.13}$$

here $N_p : N_s$ is the turns ratio of the primary and secondary windings, i_p is the primary current, and i_s the secondary current. The other state variable in the system is the capacitor voltage v. Therefore, the form of the state variables is

$$X = \begin{pmatrix} \rho \\ v \end{pmatrix} \tag{3.14}$$

in this converter, namely $x_1 = \rho$ and $x_2 = v$ in Equation 3.14.

When the system is in switching mode $(S,D) = (\text{on,off})$, i_p goes up and $i_s = 0$, so ρ goes up. When the system is in switching mode $(S,D) = (\text{off,on})$, i_s goes down and $i_p = 0$, so ρ goes down. When the system in switching mode $(S,D) = (\text{off,off})$, i_p and i_s are both equal to 0, so $\rho = 0$. Therefore, the waveform of ρ can explain the switching states and we use these characteristics of the system to obtain the main switching blocks of voltage-mode-controlled flyback converter as shown in Figure 3.15.

To obtain the bifurcation phenomena, group 1 parameters are chosen and are shown in Table 3.2. Therefore, the bifurcation diagram with controlled period T as

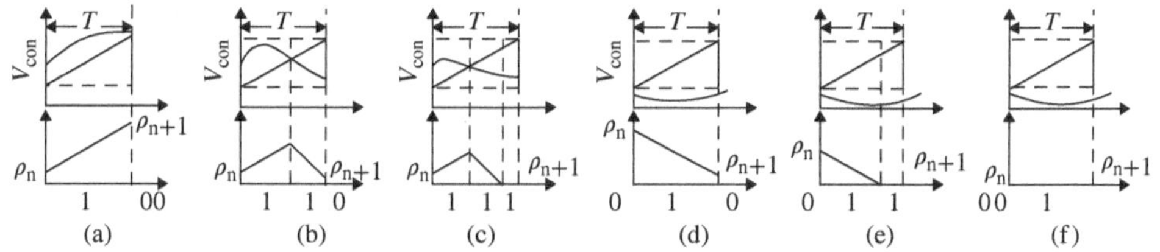

Figure 3.15 Main switching blocks of a voltage-mode-controlled flyback converter for (a) $(001)_2 = 1$, (b) $(010)_2 = 2$, (c) $(011)_2 = 3$, (d) $(100)_2 = 4$, (e) $(110)_2 = 6$, and (f) $(111)_2 = 7$

Table 3.2 The group 1 parameters of a voltage-mode-controlled flyback converter

Circuit parameters	Nominal values
Input voltage E	10 V
Capacitor C	220 μF
Load resistance R	5 Ω
Primary winding inductance L_m	1 mH
Turns ratio $N_p : N_s$	3 : 2
Gain k	1
Reference voltage V_{ref}	7 V
Upper threshold voltage V_U	8 V
Lower threshold voltage V_L	6 V
Initial value of ρ	0 AN
Initial value of v	1.3 V
Controlled period T	(200 μs, 500 μs)

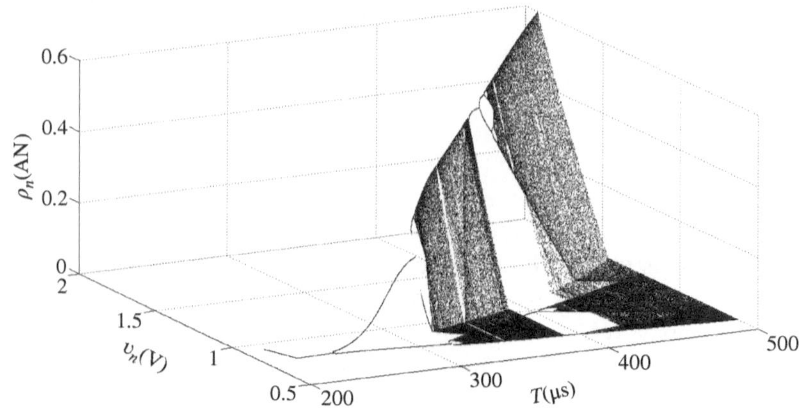

Figure 3.16 Bifurcation diagram of a voltage-mode-controlled flyback converter using group 1 parameters

Table 3.3 Duplicate symbolic sequence description of a voltage-mode-controlled flyback converter using group 1 parameters

Controlled period T (μs)	S	O	L	P_S	P_O	P_L
(200,222.5)	$(6^0)_\infty$	$(6)_\infty$	$(0)_\infty$	1	1	1
[222.5,247.9)	$(7^0)_\infty$	$(7)_\infty$	$(0)_\infty$	1	1	1
[247.9,248.2)	$(7^07^1)_\infty$	$(7)_\infty$	$(01)_\infty$	2	1	2
[248.2,320.7)	$(6^07^2)_\infty$	$(67)_\infty$	$(02)_\infty$	2	2	2
[320.7,324.3)	$(3^06^{22}7^56^{24})_\infty$	$(3676)_\infty$	$[0(22)5(24)]_\infty$	4	4	4
[327,351.2)	(∞)	(∞)	(∞)	$+\infty$	$+\infty$	$+\infty$
[389.8,390.1)	(∞)	$(376)_\infty$	(∞)	$+\infty$	3	$+\infty$
[390.1,403.9)	$(3^07^36^6)_\infty$	$(376)_\infty$	$(036)_\infty$	3	3	3
[403.9,407.4)	$(3^07^36^63^07^46^6)_\infty$	$(376)_\infty$	$(036046)_\infty$	6	3	6

the bifurcation parameter can be obtained through stroboscopic sampling, as shown in Figure 3.16. And then, using our method of duplicate symbolic sequence, we obtain some results of duplicate symbolic sequence as shown in Table 3.3.

Thus the bifurcation behaviors of a voltage-mode-controlled flyback converter using group 1 parameters are analyzed as follows.

1. When $T=222.5$ μs, $S_1=(6°)_\infty \neq S_2=(7°)_\infty$, $O_1=(6)_\infty \neq O_2=(7)_\infty$, $L_1=L_2=(0)_\infty$, $P_{S1}=P_{S2}=1$, $P_{O1}=P_{O2}=1$, and $P_{L1}=P_{L2}=1$, it explains that the system exhibits border collision bifurcation and jumps from period-1 orbit to the other period-1 orbit.

2. When $T = 247.9\,\mu s$, $S_1 = (7^{\circ})_{\infty} \neq S_2 = (7^{\circ}7^{1})_{\infty}$, $O_1 = O_2 = (7)_{\infty}$, $L_1 = (0)_{\infty} \neq L_2 = (01)_{\infty}$, $2P_{S1} = P_{S2} = 2$, $P_{O1} = P_{O2} = 1$, and $2P_{L1} = P_{L2} = 2$, it explains that the system exhibits period-doubling bifurcation and forks from period-1 orbit to period-2 orbit.
3. When $T = 248.2\,\mu s$, $S_1 = (7^{\circ}7^{2})_{\infty} \neq S_2 = (6^{\circ}7^{1})_{\infty}$, $O_1 = (7)_{\infty} \neq O_2 = (67)_{\infty}$, $L_1 = (01)_{\infty} \neq L_2 = (02)_{\infty}$, $P_{S1} = P_{S2} = 2$, $2P_{O1} = P_{O2} = 2$, and $P_{L1} = P_{L2} = 2$, it explains that the system exhibits border collision bifurcation and jumps from period-2 orbit to another period-2 orbit.
4. When $T = 320.7\,\mu s$, $S_1 = (6^{\circ}7^{2})_{\infty} \neq S_2 = (3^{\circ}6^{22}7^{5}6^{24})_{\infty}$, $O_1 = (67)_{\infty} \neq O_2 = (3676)_{\infty}$, $L_1 = (02)_{\infty} \neq L_2 = [0(22)5(24)]_{\infty}$, $2P_{S1} = P_{S2} = 4$, $2P_{O1} = P_{O2} = 4$, and $2P_{L1} = P_{L2} = 4$, it explains that the system exhibits border collision bifurcation and jumps from period-2 orbit to period-4 orbit.
5. When $T = 390.1\,\mu s$, $S_1 \neq S_2$, $P_{S1} = +\infty \gg P_{S2} = 3 O_1 = O_2 = (376)_{\infty}$, $P_{O1} = P_{O2} = 3$, $L_1 \neq L_2$, and $P_{L1} = +\infty \gg P_{L2} = 3$, it explains that the system exhibits saddle node bifurcation and forks from intermittent chaotic orbit to period-3 orbit.
6. When $T = 403.9\,\mu s$, $S_1 = (3^{\circ}7^{3}6^{6})_{\infty} \neq S_2 = (3^{\circ}7^{3}6^{6}3^{\circ}7^{4}6^{6})_{\infty}$, $O_1 = O_2 = (376)_{\infty}$, $L_1 = (036)_{\infty} \neq L_2 = (036046)_{\infty}$, $2P_{S1} = P_{S2} = 6$, $P_{O1} = P_{O2} = 3$, and $2P_{L1} = P_{L2} = 6$, it explains that the system exhibits period-doubling bifurcation and forks from period-3 orbit to period-6 orbit.

Using group 1 parameters, we did not obtain the example of Hopf bifurcation. Therefore, we choose the group 2 parameters as shown in Table 3.4 to analyze Hopf bifurcation behavior, and then the bifurcation diagram and the critical points detected by the duplicate symbolic sequence are shown in Figure 3.17 and Table 3.5, respectively. From Table 3.5, there exists Hopf bifurcation at one point between $T = 70.04\,\mu s$

Table 3.4 The group 2 parameters of a voltage-mode-controlled flyback converter

Circuit parameters	Nominal values
Input voltage E	5 V
Capacitor C	10 μF
Load resistance R	15 Ω
Primary winding inductance L_m	1 mH
Turns ratio $N_p : N_s$	3 : 2
Gain k	1
Reference voltage V_{ref}	7 V
Upper threshold voltage V_U	8 V
Lower threshold voltage V_L	0 V
Initial value of ρ	0 AN
Initial value of v	0 V
Controlled period T	(50 μs, 200 μs)

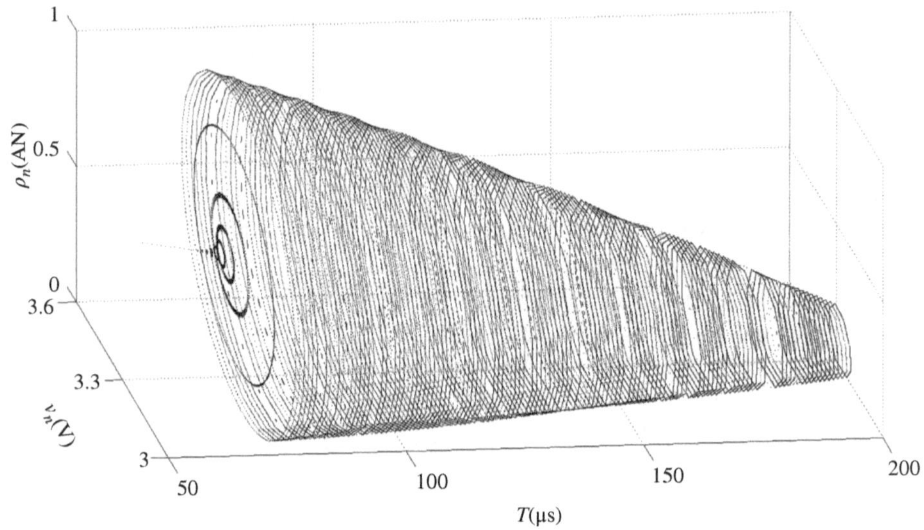

Figure 3.17 Bifurcation diagram of a voltage-mode-controlled flyback converter using group 2 parameters

Table 3.5 Duplicate symbolic sequence description of a voltage-mode-controlled flyback converter using group 2 parameters

Controlled period T (μs)	S	O	L	P_S	P_O	P_L
70.04	$(6°)\infty$	$(6)\infty$	$(0)\infty$	1	1	1
70.05	∞	$(6)\infty$	∞	$+\infty$	1	$+\infty$

and $T = 70.05$ μs, because $S_1 = (6°)_\infty \neq S_2 = \infty$, $O_1 = O_2 = (6)_\infty$, $L_1 = (0)_\infty \neq L_2 = \infty$, $P_{S1} = 1 \ll P_{S2} = +\infty$, $P_{O1} = P_{O2} = 1$, and $P_{L1} = 1 \ll P_{L2} = +\infty$.

Now we use the complexity and adopt the group 1 parameters to detect the bifurcation behavior. The complexity diagrams are shown in Figure 3.18. Moreover, we obtain the following conclusions.

1. When $T = 222.5$ μs, C remains unchanged and C_W changes, it explains that the system exhibits border collision bifurcation and jumps from a period-n orbit to another period-n orbit.
2. When $T = 247.9$ μs, C increases by only one and C_W remains unchanged, it explains that the system exhibits period-doubling bifurcation.
3. When $T = 248.2$ μs, C remains unchanged and C_W changes, it explains that the system exhibits border collision bifurcation and jumps from a period-n orbit to another period-n orbit.

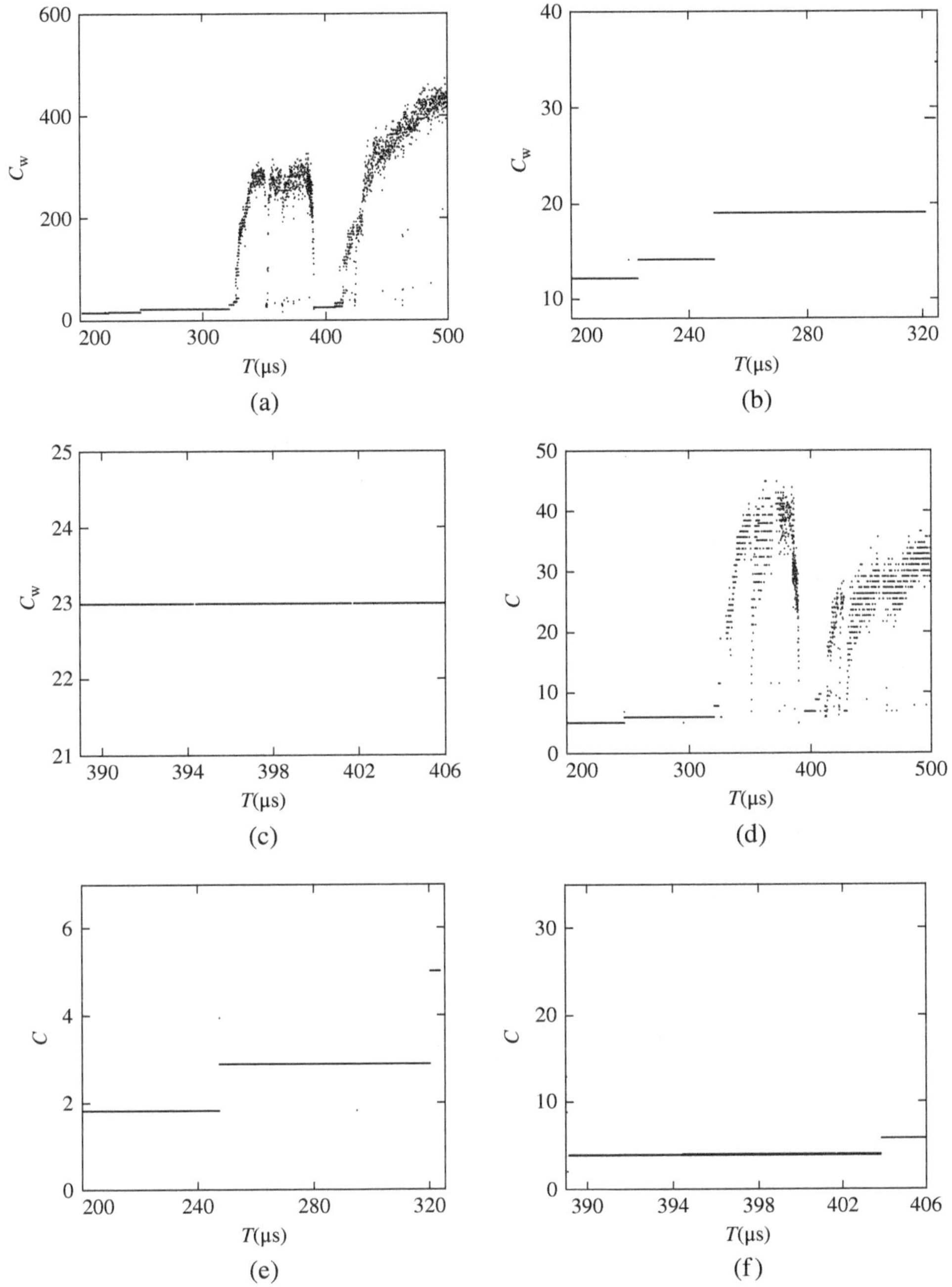

Figure 3.18 Complexity diagrams of a voltage-mode-controlled flyback converter using group 1 parameters for (a) weight complexity, (b) enlargement of (a), (c) enlargement of (a), (d) Lempel–Ziv complexity, (e) enlargement of (d), and (f) enlargement of (d)

4. When $T=320.7\,\mu s$, C and C_W both change and their increments are both less, it explains that the system exhibits border collision bifurcation.
5. When $T=390.1\,\mu s$, C_W remains unchanged but the decrement of C is more, it explains that the system exhibits saddle node bifurcation.
6. When $T=403.9\,\mu s$, the increment of C is less and C_W remains unchanged, it explains that the system exhibits period-doubling bifurcation.

3.7 Conclusions

The method of symbolic complexity was presented to analyze the complex nonlinear phenomena of DC–DC switching converters in this chapter. We introduced the Lempel–Ziv complexity first to explain the complexity of DC–DC converters. Moreover, we proposed a new concept of weight complexity in terms of DC–DC switching converters. The concepts of main switching block and main symbolic sequence in DC–DC switching converters were presented. The quantization values of the system's complexity in one switching period were obtained using the main switching block. The duplicate symbolic sequence was also proposed to detect all of the main bifurcation types in DC–DC converters.

References

[1] Giaouris, D., Banerjee, S., Imrayed, O. *et al.* (2012) Complex interaction between tori and onset of three-frequency quasi-periodicity in a current mode controlled boost converter. *IEEE Trans. Circuits Syst. Regul.Pap.*, **59**(1), 207–214.

[2] Zhou, J.B., Meng, G., Chen, J.Y. *et al.* (2009) Bifurcation analysis of Ultra short self-acting gas journal bearings for MEMS. *IEEE Trans. Ind. Electron.*, **56**(8), 3188–3194.

[3] Wang, C.S., Covic, G.A. and Stielau, O.H. (2004) Power transfer capability and bifurcation phenomena of loosely coupled inductive power transfer systems. *IEEE Trans. Ind. Electron.*, **51**(1), 148–157.

[4] Sallán, J., Villa, J.L., Llombart, A. *et al.* (2009) Optimal design of ICPT systems applied to electric vehicle battery charge. *IEEE Trans. Ind. Electron.*, **56**(6), 2140–2149.

[5] Tse, C.K. (2003) *Complex Behavior of Switching Power Converters*, CRC Press, Boca Raton, FL.

[6] Duran, M.J., Salas, F. and Arahal, M.R. (2008) Bifurcation analysis of five-phase induction motor drives with third harmonic injection. *IEEE Trans. Ind. Electron.*, **55**(5), 2006–2014.

[7] Pantic, Z., Bai, S. and Lukic, S.M. (2011) ZCS LLC-compensated resonant inverter for inductive-power-transfer application. *IEEE Trans. Ind. Electron.*, **58**(8), 3500–3510.

[8] Orabi, M. and Ninomiya, T. (2003) Nonlinear dynamics of power-factor-correction converter. *IEEE Trans. Ind. Electron.*, **50**(6), 1116–1125.

[9] Aroudi, A.E., Orabi, M., Haroun, R. *et al.* (2011) Asymptotic slow-scale stability boundary of PFC AC-DC power converters: theoretical prediction and experimental validation. *IEEE Trans. Ind. Electron.*, **58**(8), 3448–3460.

[10] Wang, J., Bao, B., Xu, J.R. *et al.* (2013) Dynamical effects of equivalent series resistance of output capacitor in constant on-time controlled buck converter. *IEEE Trans. Ind. Electron*, **60**(5), 1759–1768, http://ieeexplore.ieee.org/stamp/stamp.jsp?tp=&arnum-ber=6169989 (accessed 7 June 2014).

[11] Xie, F., Yang, R. and Zhang, B. (2011) Bifurcation and border collision analysis of voltage-mode-controlled flyback converter based on total ampere-turns. *IEEE Trans. Circuits Syst. Regul. Pap.*, **58**(9), 2269–2280.
[12] Dranga, O., Buti, B. and Nagy, I. (2003) Stability analysis of a feedback-controlled resonant DC-DC converter. *IEEE Trans. Ind. Electron.*, **50**(1), 141–152.
[13] Rodriguez, E., Aroudi, A.E., Guinjoan, F. *et al.* (2012) A ripple-based design-oriented approach for predicting fast-scale instability in DC-DC switching power supplies. *IEEE Trans. Circuits Syst. Regul. Pap.*, **59**(1), 215–227.
[14] Zhusubaliyev, Z.T., Mosekilde, E. and Yanochkina, O. (2011) Torus-bifurcation mechanisms in a DC/DC converter with pulse width-modulated control. *IEEE Trans. Power Electron.*, **26**(4), 1270–1279.
[15] Tse, K.K., Ng, R.W.M., Chung, H.S.H. *et al.* (2003) An evaluation of the spectral characteristics of switching converters with chaotic carrier-frequency modulation. *IEEE Trans. Ind. Electron*, **50**(1), 171–182.
[16] Yang, R., Zhang, B., Zhao, S.B. *et al.* (2010) Arithmetic complexity of discrete map of converter based on symbol time series. *Acta Phys. Sinica*, **61**(11), 110504.
[17] Wang, X.M., Zhang, B. and Qiu, D.Y. (2008) Symbolic time series characterization and block entropy analysis of DC-DC converters. *Acta Phys. Sinica*, **57**(10), 6112–6119.
[18] Wang, X.M., Zhang, B. and Qiu, D.Y. (2011) The quantitative characterization of symbolic series of a boost converter. *IEEE Trans. Power Electron.*, **26**(8), 2101–2105.
[19] Dai, D. and Tse, C.K. (2004) Symbolic analysis of bifurcation in switching power converters: a novel method for detecting border collision. International Workshop on Nonlinear Circuit and Signal Processing (NCSP'04), Hawaii, pp. 5–7.
[20] Dai, D., Tse, C.K. and Ma, X. (2005) Symbolic analysis of switching systems: Application to bifurcation analysis of DC-DC switching converters. *IEEE Trans. Circuits Syst. Reg. Pap.*, **52**(8), 1632–1643.
[21] Tse, C.K. and Dai, D. (2005) Symbolic analysis of bifurcation in switching power converters: a practical alternative viewpoint of border collision. *Int. J. Bifurcation Chaos*, **15**(7), 2263–2270.
[22] Li, M. and Vitányi, P.M.B. (2008) *An Introduction to Kolmogorov Complexity and its Applications*, Springer Science Business Media, LLC.
[23] Lempel, A. and Ziv, J. (1976) On the complexity of finite sequence. *IEEE Trans. Inf. Theory*, **22**(1), 75–81.
[24] Zhang, X.S., Zhu, Y.S., Thakor, N.V. *et al.* (1999) Detecting ventricular tachycardia and fibrillation by complexity measure. *IEEE Trans. Biomed. Eng.*, **46**(5), 548–555.
[25] Xie, F., Yang, R. and Zhang, B. (2012) Analysis of weight Lempel-Ziv complexity in piecewise smooth systems of DC-DC switching converters. *Acta Phys. Sinica*, **59**(6), 3756–3762.
[26] Xie, F., Zhang, B., Yang, R. and Iu, H.H.C. (2013) Detecting bifurcation types and characterizing stability in DC–DC switching converters by duplicate symbolic sequence and weight complexity. *IEEE Trans. Ind. Electron.*, **60**(8), 3145–3156.

4

Invariant Probability Distribution of DC–DC Converters

4.1 Introduction

This chapter is concerned with applying probability analysis to chaotic mapping of DC–DC converters. A computation method for the invariant probability distribution of a chaotic mapping is proposed by using the eigenvector method, which is to facilitate the accurate design of the DC–DC converter parameters. Moreover, the power spectral density of the input to the DC–DC converter and the average frequency of switching are deduced. Finally, some application examples are given to illustrate the effectiveness of the proposed method.

It is known that chaotic motion is an unstable, aperiodic behavior within a bounded area. Moreover its long-term behavior shows random-like characteristics, which can be studied using probability theory. The invariant probability distribution, providing us with a satisfactory description of the chaos motion, is a basic and important characteristic of chaos. For a DC–DC converter, a one-dimensional mapping can be derived under some reasonable assumptions, which can then be used to analyze the chaotic behavior of the DC–DC converter. There have been several methods proposed to calculate the invariant probability distribution of the chaotic mapping of the DC–DC converter. However, these methods have their own drawbacks. For instance, the method presented in [1] is difficult to realize by computer due to the immense increase in the calculation complexity when the iteration of the mapping increases slightly. Moreover, this method does not require the mapping to have a finite number of Markov partitions [2]. The method in [3] uses the Frobenius–Perron operator equation to calculate the invariant probability distribution. But it is well known that only a few Frobenius–Perron operator equations of chaotic mappings may be solved analytically, thus this method can only be applied in a few special cases.

Chaos Analysis and Chaotic EMI Suppression of DC-DC Converters, First Edition. Bo Zhang and Xuemei Wang.

In this chapter, a boost converter operating in a chaotic mode is described by a one-dimensional mapping. And the invariant probability distribution of the chaotic mapping is then calculated using the eigenvector method. The comparison of the invariant probability distribution of the chaotic mapping with its phase portrait and its bifurcation diagram shows that the method is appropriate to calculate the invariant probability distribution of the chaotic mapping of the DC–DC converter. Furthermore, the calculation results can also be used to estimate the power spectral density of the input, to calculate the average switching frequency of DC–DC converters, and to accurately design the system parameters. Finally, simulation examples will be given to illustrate the effectiveness of the method.

4.2 Invariant Probability Distribution of Chaotic Map

Chaos is a kind of unstable behavior in a bounded area. Its long-term behavior shows random like characteristics. Supposing that we have obtained a series of chaotic track points $\{\theta_n, n=0, 1, \ldots\}$, one can ask a question: how are these points distributed on a certain interval? Thus, it is possible to characterize it with probability theory by using the probability distribution $\rho(x)$ of the chaotic mapping. The answer defines the density of points

$$\rho(\theta) = \lim_{\Delta\theta\to 0} \lim_{n\to\infty} \frac{\text{number of } \theta_n \in (\theta,\ \theta + \Delta\theta)}{n\Delta\theta} \tag{4.1}$$

The distribution of density must be invariant, that is, the number of orbit points of chaotic maps remains invariant under the mapping.

For some simple cases, such as the parabola mapping and piecewise linear maps, $\rho(\theta)$ can be calculated exactly. But for general cases, calculating $\rho(x)$ requires employment of the the Perron–Frobenious equation to get a numerical solution. The Perron–Frobenious equation is based on "points conservation" [4, 5]. Figure 4.1 shows a nonlinear function, where y has two inverse images x_1 and x_2, namely, $y=f(x_1)=f(x_2)$.

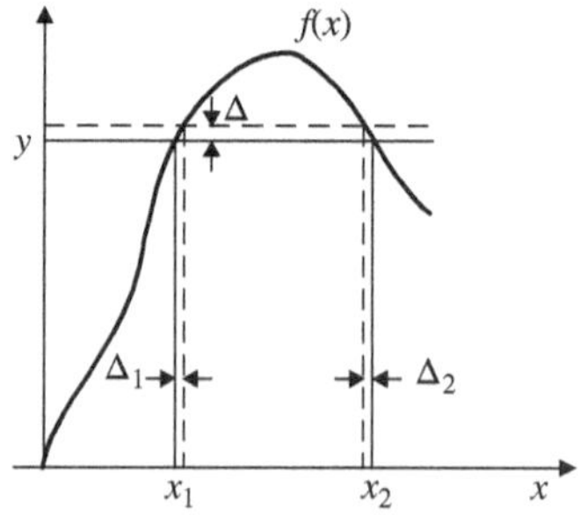

Figure 4.1 Mapping of a nonlinear function

Denote the small cell of the neighborhoods of x_1, x_2, and y as Δ_1, Δ_2, and Δ, respectively, and the corresponding probability densities as $\rho(x_1)$, $\rho(x_2)$, and $\rho(y)$. According to the conservation law of quantity [4, 5], one has

$$\rho(y)\Delta = \rho(x_1)\Delta_1 + \rho(x_2)\Delta_2 \tag{4.2}$$

When the small cells are thin enough, the relationship among Δ, Δ_1, and Δ_2 is determined by the derivative

$$\Delta_i = \frac{\Delta}{|f'(x_i)|}$$

So, Equation 4.2 can be recast as

$$\rho(y) = \frac{\rho(x_1)}{|f'(x_1)|} + \frac{\rho(x_2)}{|f'(x_2)|} \tag{4.3}$$

where, $f(x_1) = \Delta/\Delta_1$ and $f(x_2) = \Delta/\Delta_2$. If $f(x)$ has multiple inverse images (more than 2), namely, there exist $x_i = f^{-1}(y)$, $i > 2$. So Equation 4.3 can be denoted as

$$\rho(y) = \sum_{\{x_i = f^{-1}(y)\}} \frac{\rho(x_i)}{|f'(x_i)|} \tag{4.4}$$

This is the so-called Perron–Frobenious equation, on which the calculation of the invariant probability distribution using the eigenvector method can be based.

For the logistic map Equation 2.1, one has

$$|f'(x_i)| = 4|x|$$

Here, $x = \pm\sqrt{(1-y)/2}$. So, the Perron–Frobenious equation can be specifically written as

$$\rho(y) = \frac{1}{\sqrt{2(1-y)}}\rho(\sqrt{(1-y)/2})$$

And its solving is

$$\rho(y) = \frac{1}{\pi\sqrt{1-x^2}} \tag{4.5}$$

The invariant probability distribution of logistic map is illustrated in Figure 4.2 using Equation 4.5.

If one solves the Perron–Frobenious equation with numerical methods, Equation 4.4 may be rewritten as an iterative equation

$$\rho_n(y) = \sum_{\{x_i = f^{-1}(y)\}} \frac{\rho_{n-1}(x_i)}{|f'(x_i)|} \tag{4.6}$$

Taking a reasonable initial probability distribution value $\rho_0(x)$ as constant, a convergence invariant probability distribution may be obtained by iterating several times.

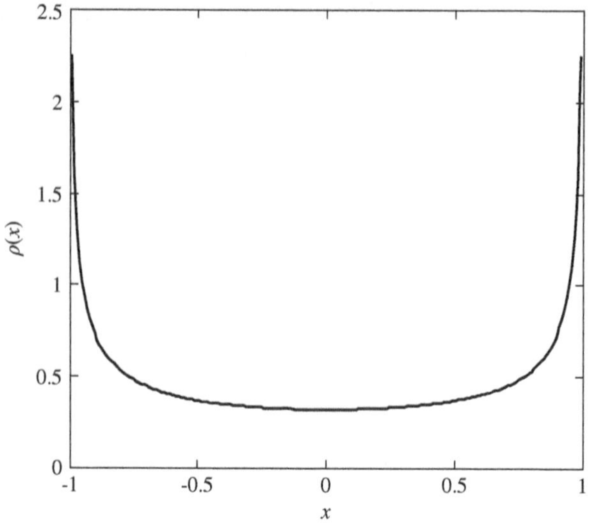

Figure 4.2 Invariant probability distribution of logistic map with $\mu = 4$

4.3 Calculating Invariant Probability Distribution of the Chaotic Discrete-Time Maps with Eigenvector Method

For a nonlinear function $f(x)$, $f: I \to I$, the interval I can be equally divided into M segments. If M is large enough, $\rho(x)$ can be regarded as "invariant" in each small interval.

Then, $\rho(x)$ can be expressed as M discrete values $\rho(x_1)$, $\rho(x_2)$, …, $\rho(x_M)$ or the vector form $\boldsymbol{R} = [\rho(x_1), \rho(x_2), \ldots, \rho(x_M)]$.

In Figure 4.3, $p_{i,j}$ is the transition probability of the jth interval, and the transition probability matrix is denoted as

$$P = \begin{vmatrix} p_{1,1} & p_{1,2} & \cdots & p_{1,M} \\ p_{2,1} & p_{2,2} & \cdots & p_{2,M} \\ \vdots & \vdots & \ddots & \vdots \\ p_{M,1} & p_{M,2} & \cdots & p_{M,M} \end{vmatrix} \tag{4.7}$$

Where, the entries can be derived by

$$\begin{cases} p_{m,j} = (x_n - x_m)/L, \\ p_{m+1,j} = (x_s - x_n)/L, \\ p_{m+2,j} = (x_c - x_s)/L, \\ p_{i,j} = 0 (1 \le i \le M, i \ne m, m+1, m+2) \end{cases} \tag{4.8}$$

Thus, calculating the transition probability matrix P is easy when $f(x)$ and M are known.

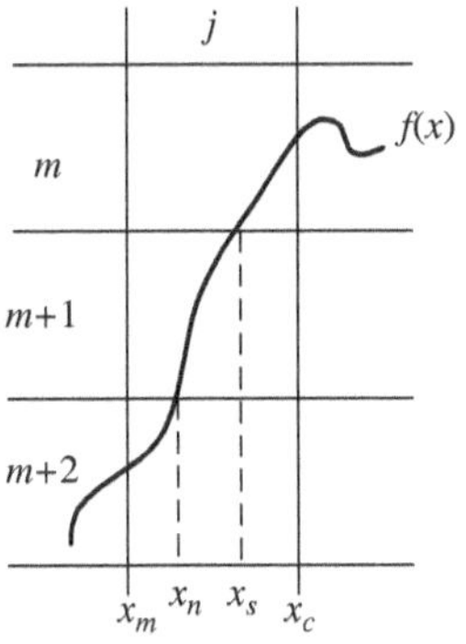

Figure 4.3 Partial sketch mapping of chaotic mapping

Definition 4.1 A probability distribution P is called *invariant* if

$$PR = R \tag{4.9}$$

where, P is a left eigenvector of R with eigenvalue 1. Thus, the calculation of the invariant probability distribution is reduced to a calculation of the eigenvector of the transition probability matrix P.

4.4 Invariant Probability Distribution of the Chaotic Mapping of the Boost Converter

The peak-current control boost converter shown in Figure 4.4 is an example to introduce the calculation of invariant probability distribution of chaotic mapping [7].

Possible inductor current i_L waveforms that may occur in continuous conduction mode (CCM) states with peak-current control are shown in Figure 4.5, where I_{ref} is the peak reference current. The converter will undergo two work modes in a single clock cycle if the duty cycle is less than 1 as shown in Figure 4.5a. If the duty cycle

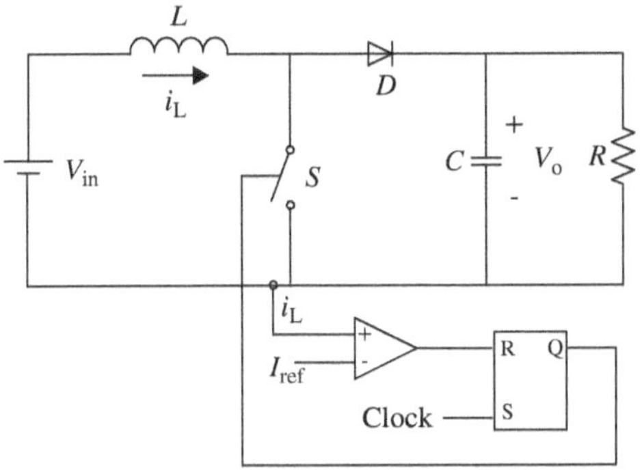

Figure 4.4 Boost converter with peak-current control

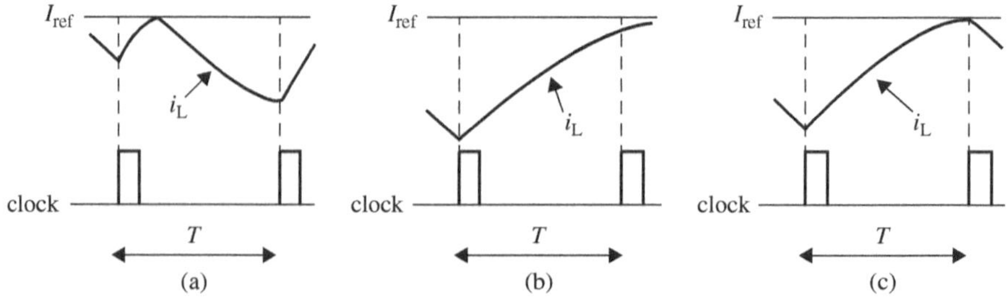

Figure 4.5 Current waveforms in CCM of peak-current control for (a) undergoing two modes, (b) undergoing one mode, and (c) boundary case

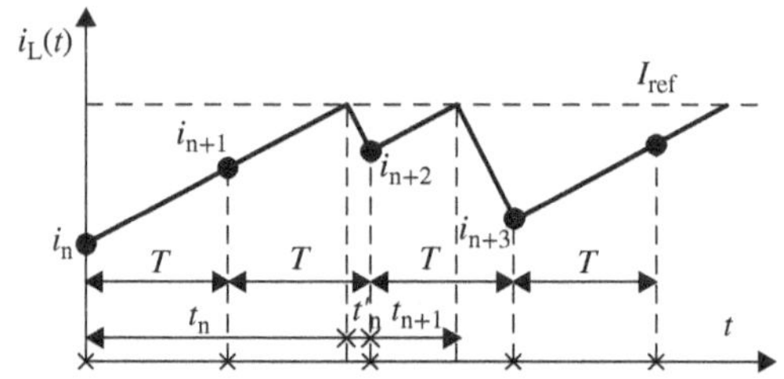

Figure 4.6 The current waveform $i_L(t)$ in a boost converter

equals to 1, the converter works only in one mode and current i_L cannot reach to I_{ref} in the entire clock cycle as shown in Figure 4.5b. If the current i_L just reaches to I_{ref} at the end of the clock cycle, this case is called a border situation as shown in Figure 4.5c. For the converters working in CCM states with peak-current control, we denote the collision boundary line

$$I_{border-1} = I_{ref} - \frac{E}{L}T \tag{4.10}$$

Let i_L at the nth switching moment be i_n, and those at the next moment be i_{n+1}. If the output capacitance of the boost converter tends to infinity or is big enough in practice, such as $v_o = V_{out}$, the inductor current can be described as one-dimensional iterative map [8, 9]

$$\begin{cases} i_{n+1} = i_n + m_1 T & i_n \le i_{border-1} \\ i_{n+1} = I_{ref}(1 + m_2/m_1) - m_2 T - i_n m_2/m_1 & i_n > i_{border-1} \end{cases} \tag{4.11}$$

as shown in Figure 4.6, where $m_1 = \frac{E}{L}$ and $m_2 = \frac{V_o - E}{L}$ are the rates of increment and decrement of the inductor current, respectively.

The duration of a switching cycle is $t_n + t'_n = T(1 + \lfloor x_n \rfloor)$. Notice that $t'_n = T(1 - (x_n \mathrm{mod} 1))$ where $x_n = t_n/T$ is the normalized in *on* interval. Recognizing that the *on* interval is given by $t_n = (I_{ref} - i_n)/c_1$, an impact map reads as [9]

$$x_{n+1} = \alpha(1 - (x_n \mathrm{mod} 1)) \tag{4.12}$$

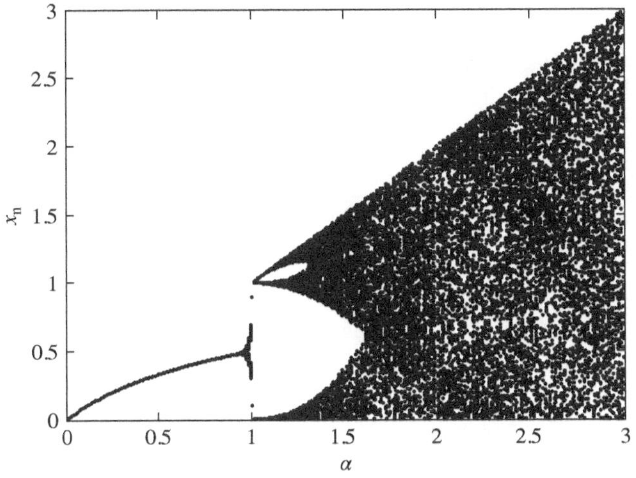

Figure 4.7 Borderline and bifurcation diagram

where $t_n = x_n T_c$, $t'_n = T_C(1 - (x_n \mathrm{mod} 1))$, $\alpha = \frac{\overline{V}_{out}}{V_{in}} - 1$, and $t_n = \frac{(I_{ref} - i_n)L}{V_I}$. The border line and bifurcation diagram of Equation 4.12 are shown in Figure 4.7. One can find that the system jumps directly from period-1 to chaos, which is called robust chaos.

For a boost converter, one has $\alpha > 0$ due to $\overline{V}_{out} > V_{in}$. It is easy to see by the Lyapunov exponent that for $\alpha > 1$ the sequence $\{x_0, x_1, \ldots, x_n, \ldots\}$ is chaotic within the range $[0, \alpha]$ [4, 5]. The mapping Equation 4.12, or its normalization, has been extensively studied, most notably by Rényi [10, 11]. Rényi shows the Frobenius–Perron equation defined in Section 4.2 that has an invariant probability distribution $\rho(x)$. Moreover, the Frobenius–Perron equation is absolutely continuous with respect to the Lebesgue measure on the interval $[0\ \alpha]$, and ergodic and asymptotically stable [11]. Due to the random-like characteristic of chaos, the eigenvector method extracted from probability theory will be introduced next and is employed here to calculate the invariant probability distribution of a chaotic mapping.

According to Equations 4.1–4.6 the eigenvector $\boldsymbol{R} = [\rho(x_1), \rho(x_2), \ldots, \rho(x_M)]$ of P, namely, the invariant probability distribution of the chaotic mapping, can be calculated by dividing the interval $[0\ \alpha]$ into M equal segments for the above mentioned boost converter. When α samples different values, the simulation results are presented as follows.

When $\alpha = 1.30$, the phase portrait of the mapping, the corresponding probability histogram diagram, and the invariant probability distribution are shown in Figures 4.8a–c respectively. Here 20 000 discrete data are utilized for the probability histogram.

From these figures, it is clear that they inosculate quite well. It is remarked that the invariant probability distribution reflects the operating status of the boost converter from a special perspective. It can be seen from Figure 4.9a that there are no orbit points in the intervals [0.13, 0.91] and [1.10, 1.15], corresponding to the zero

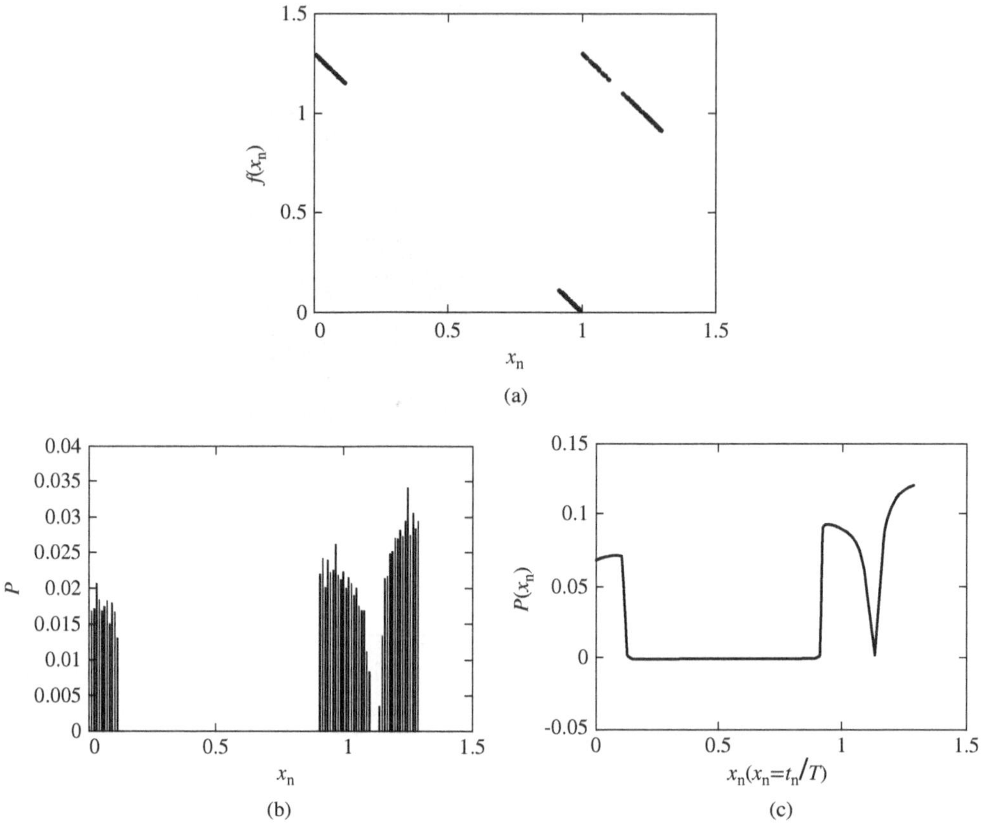

Figure 4.8 (a) The chaotic mapping, (b) probability histogram, and (c) invariant probability distribution at $\alpha = 1.30$

invariant probability distribution in those intervals. Similarly, for the cases $\alpha = 1.52$ and 2.65, the simulation results are illustrated in Figures 4.8–4.10. The simulation results illustrate the accuracy of the eigenvector method in calculating the invariant probability distribution.

4.5 Application Examples of Invariant Probability Distribution

The invariant probability distribution can be used to calculate the power spectral density of the input of the DC–DC converter, to estimate the average switching frequency, and to accurately design the parameters of the DC–DC converter. Two examples are given in the following for illustration.

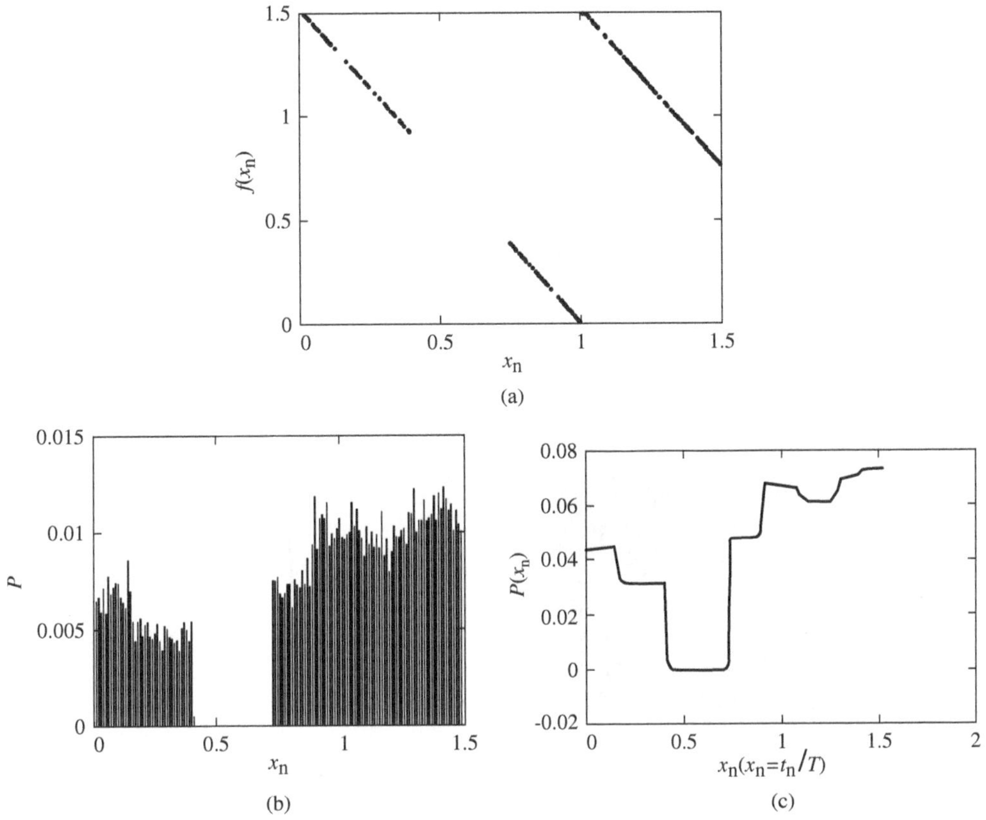

Figure 4.9 (a) The chaotic mapping, (b) probability histogram, and (c) invariant probability distribution at $\alpha = 1.52$

4.5.1 Power Spectral Density of the Input Current in a DC–DC Converters

Consider the above introduced boost converter. According to Figure 4.6, the quadratic derivative of the inductor current is shown in Figure 4.11 [7]. According to [1], the inductor current can be expressed as

$$\frac{d^2 i}{dt^2} = -\frac{\overline{V}_O}{L}\left\{\delta(t) - \delta\left[t - T_C x_1\right] + \delta\left[t - T_C(1 + \lfloor x_1 \rfloor)\right] - \delta\left[t - T_C(1 + \lfloor x_1 \rfloor + x_2)\right]\right.$$
$$\left. + \cdots - \delta\left[t - T_C\left(N - 1 + \sum_{k=1}^{N-1} \lfloor x_k \rfloor\right) + x_N\right] + \delta\left[t - T_C\left(N + \sum_{k=1}^{N} \lfloor x_k \rfloor\right)\right]\right\} \tag{4.13}$$

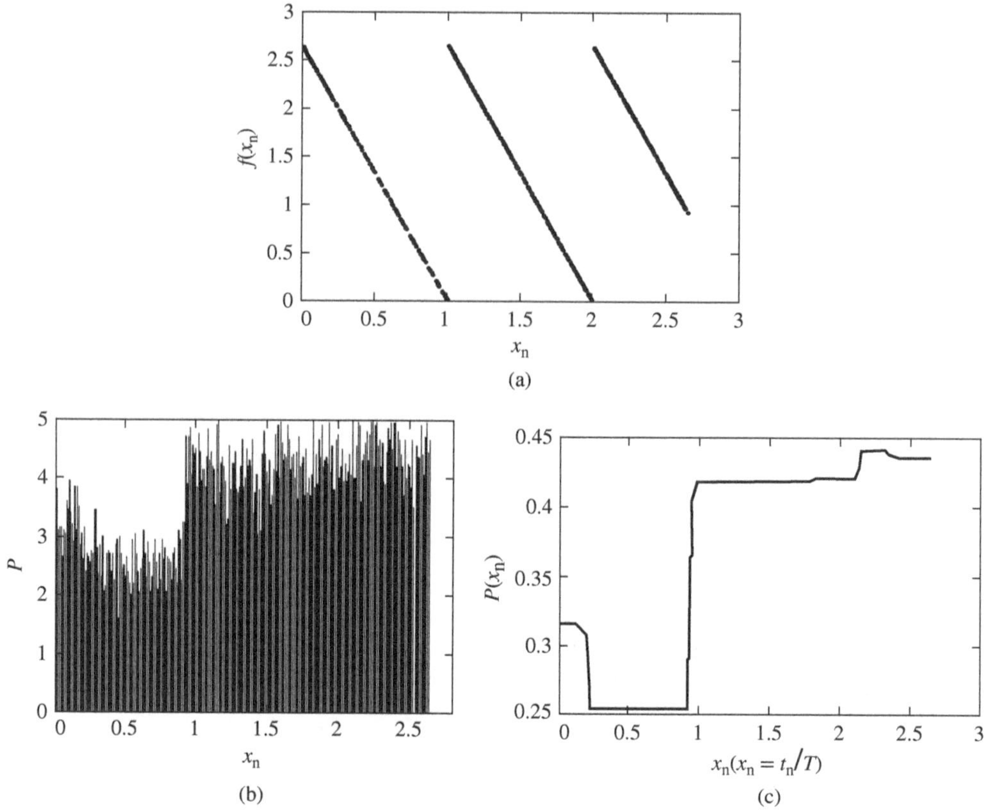

Figure 4.10 (a) The chaotic mapping, (b) probability histogram, and (c) invariant probability distribution at $\alpha = 2.65$

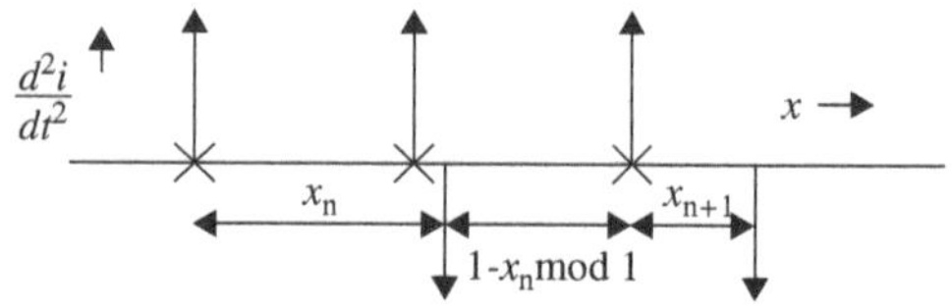

Figure 4.11 Quadratic derivative of inductor current

where $\lfloor x \rfloor$ means the round-off number. Employing the following Fourier transformation

$$g(t) \triangleq G(\omega) \Rightarrow \int_{-\infty}^{t} g(u)du \rightleftharpoons \frac{1}{j\omega}G(\omega)$$

$$g(t) \rightleftharpoons G(\omega) \Rightarrow g(t-\tau) \rightleftharpoons e^{-j\omega t}G(\omega)$$

and

$$\delta(t) \rightleftharpoons 1$$

So, the results in the Fourier transform of the inductor current are as

$$A(\omega) = -\frac{\overline{V}_O}{\omega^2 L} \lim_{N\to\infty} \frac{1}{T_N} \left[\{1 - \exp\left(-j\omega T_C x_1\right)\} + \exp\left(-j\omega T_C\left[1 + \lfloor x_1 \rfloor\right]\right) \right.$$
$$\times \{1 - \exp\left(j\omega T_C x_2\right)\} + \cdots + \exp\left(-j\omega T_C\left[N - 1 + \sum_{k=1}^{N-1} \lfloor x_k \rfloor\right]\right)$$
$$\left. \times \{1 - \exp\left(-j\omega T_C x_N\right)\}\right] \tag{4.14}$$

Then, we denote

$$J_n = \begin{cases} 0 & \text{for} \quad n = 1 \\ \sum_{k=1}^{N-1} 1 + \lfloor x_k \rfloor = n - 1 + \sum_{k=1}^{N-1} \lfloor x_k \rfloor & \text{for} \quad n > 1 \end{cases} \tag{4.15}$$

and

$$T_n = \sum_{n=1}^{N} 1 + \lfloor x_n \rfloor \tag{4.16}$$

Equation 4.14 can be rewritten as

$$A(\omega) = -\frac{\overline{V}_O}{\omega^2 L} \lim_{N\to\infty} \frac{1}{T_N} \sum_{N=1}^{n} e^{-j\omega T_C J_n} \{1 - e^{-jwT_C x_n}\} \tag{4.17}$$

The power spectral density of the inductor current is defined as $|A(\omega)|^2$. When $\omega = m\omega_c$, where ω_c is clock angular frequency, one has

$$A_m = -\frac{\overline{V}_O}{\omega^2 L} \lim_{N\to\infty} \frac{1}{T_N} \sum_{n=1}^{N} 1 - e^{-2j\pi m x_n} \tag{4.18}$$

where A_m stands for the peak values.

According to Birkhoff's ergodic theory [12], a mapping f, which is of invariant probability distribution, satisfies the following relationship,

$$\lim_{N\to\infty} \frac{1}{N} \sum_{n=1}^{N} \phi(f^{n-1}(x)) = \int_0^{\alpha} \phi(y)\rho(y)dy \tag{4.19}$$

Thus, $\tilde{A}_m \triangleq |A(\omega)|^2$ can be expressed by the invariant probability distribution $\rho(x)$ as

$$\tilde{A}_m = |A_m|^2 = \left[-\frac{\overline{V}_O}{m^2\omega_c^2 L\langle T\rangle}\right]^2 \times \left[\left(\int_0^{\alpha} \cos 2\pi m x dx - 1\right)^2 + \left(\int_0^{\alpha} \sin 2\pi m x dx\right)^2\right] \tag{4.20}$$

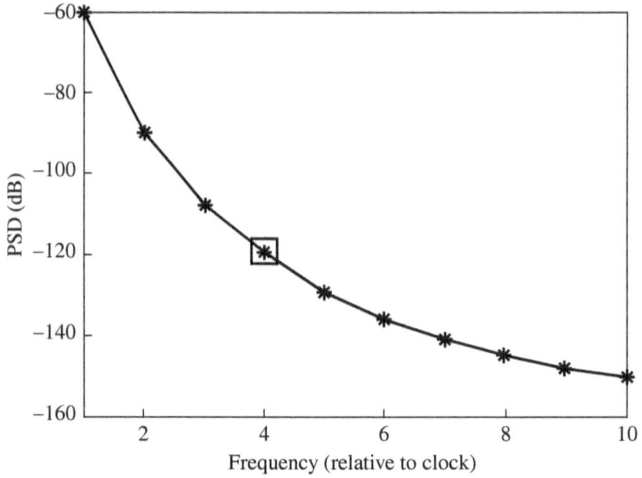

Figure 4.12 Comparison between Equation 4.18 ploted in "+" and Equation 4.20 plotted in "*."

where

$$\langle T\rangle = \lim_{n\to\infty}\frac{T_N}{N} = T_C\left(1+\lim_{N\to\infty}\frac{1}{N}\lfloor x_n\rfloor\right) \tag{4.21}$$

A comparison of the power spectral densities calculated by Equation 4.18 and by Equation 4.20 is illustrated in Figures 4.12 and 4.13. Figures 4.12 and 4.13 show that they both have almost the same accuracy, but the calculation by using the invariant probability distribution takes a much shorter time. Because Equation 4.18 includes an exponent operation, it needs to be calculated N times, and requires $N\to\infty$. While Equation 4.20 just needs a one-time calculation since the invariant probability distribution is known.

4.5.2 *Average Switching Frequency*

Chaos control in a DC–DC converter cannot only reduce electromagnetic interference (EMI) of the DC–DC converter, but also reduce the average switching frequency. Moreover, the average switching frequency is very important for reducing the switching loss and increasing the stability of a DC–DC converter. And it can be calculated with the invariant probability distribution.

For a boost converter as shown in Figure 4.4, if it operates in order, one can think that the total increment of the inductor current $\Delta_{i+(\text{total})}$ is equal to the total decrement of the inductor current $\Delta_{i-(\text{total})}$ for a relatively long time, namely, $\Delta_{i+(\text{total})} = \Delta_{i-(\text{total})}$ as shown in Figure 4.14. From Figure 4.14, the total time

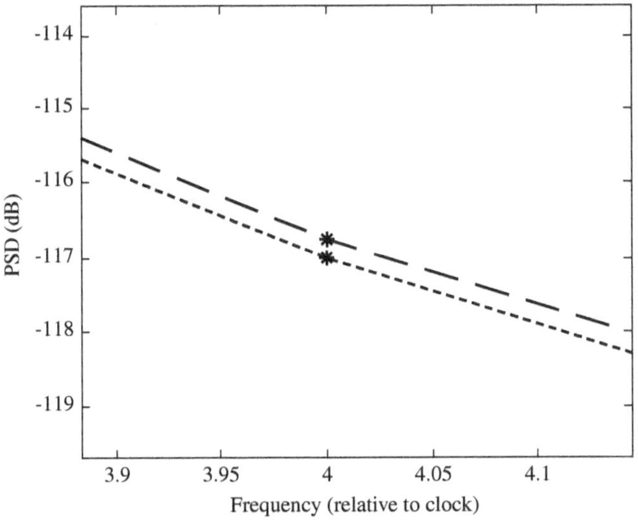

Figure 4.13 Enlargement of Figure 4.12

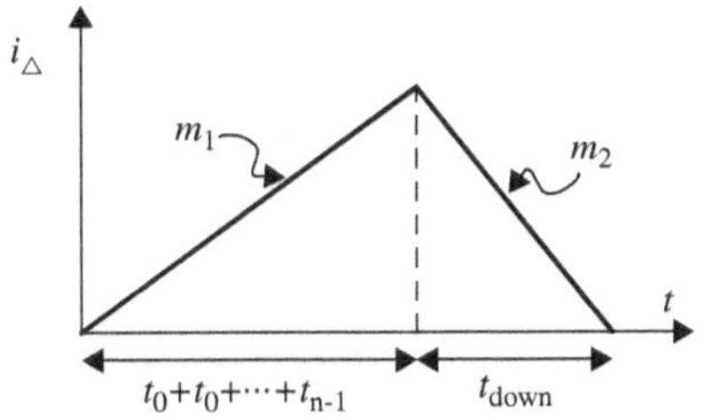

Figure 4.14 Equivalent figure

corresponding to the increasing inductor current is $(t_0 + t_1 + \ldots + t_{N-1})$, then the total time corresponding to the decreasing inductor current t_{down} can be obtained by

$$(t_0 + t_1 + \cdots + t_{N-1})m_1 = t_{down}m_2 \tag{4.22}$$

Then, t_{down} can be obtained from Equation 4.22 as

$$t_{down} = \frac{m_1}{m_2}(t_0 + t_1 + \cdots + t_{N-1}) = \frac{1}{\alpha}(t_0 + t_1 + \cdots + t_{N-1}) \tag{4.23}$$

and the total time of N-times switching is

$$T_N = \left(1 + \frac{1}{\alpha}\right)(t_0 + t_1 + \cdots + t_{N-1}) = \left(1 + \frac{1}{\alpha}\right)(x_0 + x_1 + \cdots + x_{N-1})T_C \tag{4.24}$$

Thus, the total number of clock cycle denoted by L is

$$L = \frac{T_N}{T_C} = \left(1 + \frac{1}{\alpha}\right)(x_0 + x_1 + \cdots + x_{N-1}) \tag{4.25}$$

and the total times of switching is N.

The average switching frequency is defined as [13],

$$\langle s \rangle = \lim_{N\to\infty} \frac{N}{L} = \lim_{N\to\infty} \frac{N}{\left(1 + \frac{1}{\alpha}\right)(x_0 + x_1 + \cdots + x_{N-1})} = \frac{1}{\left(1 + \frac{1}{\alpha}\right)\left(\int_0^{\alpha} \rho(x)\, x dx\right)} \tag{4.26}$$

To simplify the analysis, let the integer α be larger than 1. By chaotic mapping, it is easy to find that $\rho(x) = 1/\alpha$ for the integer $\alpha \geq 1$. Then the average switching frequency can be obtained as

$$\langle s \rangle = \frac{2}{1 + \alpha} \tag{4.27}$$

From Equation 4.27, it is obvious that $<s> = 1$ at $\alpha = 1$. This implies that the boost converter runs periodically and $<s> <1$ at $\alpha > 1$. The boost converter will operate in a chaotic mode at $\alpha > 1$. In this case, the boost converter has a low average switching frequency and low switching loss. Furthermore, the average switching frequency decreases as α increases.

4.5.3 Parameter Design with Invariant Probability Distribution

To design a DC–DC converter, such as the boost converter shown in Figure 4.4, one needs to know the value of reference current I_{ref}. Generally speaking, the values of the input voltage and the output voltage are known conditions.

In order to use the map Equation 4.12, we need to know the value of $\alpha = \overline{V}_{\text{out}}/V_{\text{in}} - 1$. Hence there remains the problem of $\overline{V}_{\text{out}}$, the mean output voltage. We obtain an estimated $\overline{V}_{\text{out}}$ by assuming that switch S is closed on average for a time $\overline{D}T_C$ every clock cycle and $\overline{V}_{\text{out}}$ is constant. Here, $\overline{D}$ is to be interpreted as the mean duty cycle of switch S. The average output voltage is

$$\overline{V}_{\text{out}} = R(1 - \overline{D})\left(I_{\text{ref}} - \frac{\overline{\Delta}_i}{2}\right) \tag{4.28}$$

where R is the load resistance of the converter.

Assume that the current through the diode is zero (for a fraction $\overline{D}$ of the time) or $I_{\text{ref}} - \overline{\Delta}_i/2$ (for a fraction $1 - \overline{D}$ of the time). When S is closed, i rises at a rate of V_{in}/L for a time $\overline{D}T_C$, on average. When S is open, i falls at a rate of $(V_{\text{in}} - V_{\text{out}})/L$ for a time $(1 - \overline{D})T_C$. The current rise $\overline{\Delta}_{i+}$ and fall $\overline{\Delta}_{i-}$ are

$$\overline{\Delta}_{i+} = \frac{V_I}{L}\overline{D}T_C \tag{4.29}$$

$$\overline{\Delta}_{i-} = \frac{\overline{V}_{\text{out}} - V_{\text{in}}}{L}(1 - \overline{D})T_C \tag{4.30}$$

And they have the relationship between Equations 4.29 and 4.30

$$\overline{\Delta}_{i+} = \overline{\Delta}_{i-} = \overline{\Delta}_i \tag{4.31}$$

Eliminating $\overline{\Delta}_{i+}$, $\overline{\Delta}_{i-}$, $\overline{\Delta}_i$, and $\overline{D}$ from Equations 4.28 and 4.29, we obtain

$$\overline{V}_{\text{out}}^3 + \overline{V}_{\text{out}}\left(\frac{V_{\text{in}}T_C}{2L} - I_{\text{ref}}\right)RV_{\text{in}} - \frac{RT_C V_{\text{in}}^3}{2L} = 0 \tag{4.32}$$

Equation 4.32 can be solved for $\overline{V}_{\text{out}}$ by selecting the real root $\overline{V}_{\text{out}} > V_{\text{in}}$. Hence, α can be found. Now we completely get the discrete-time map of the boost converter, and I_{ref} can be calculated from this formula.

By the invariant probability distribution, one can accurately design the parameters for a chaotic DC–DC converter. To simplify the calculation, α samples integers from 2 to 10. Because when α larger than 1, the corresponding invariant probability distribution is $1/\alpha$.

Denote the quantity of electric charge through the diode D at the nth time as $Q(x_n)$. Referring to the Figure 4.6 and using the physical definition of quantity of electric charge, one has

$$Q(x_n) = \left(I_{\text{ref}} - \frac{m_2\,(1 + \lfloor x_n \rfloor - x_n)\,T_c}{2}\right)(1 + \lfloor x_n \rfloor - x_n)T_c \tag{4.33}$$

Using Birkhoff's ergodic theory and the invariant probability distribution, one can get

$$\langle T \rangle = \lim_{N\to\infty}\frac{1}{N}\sum_{n=0}^{N-1} T_c(1 + \lfloor x_n \rfloor) = T_c\int_0^{\alpha}(1 + \lfloor x \rfloor)\rho(x)dx = \frac{\alpha+1}{2}T_c \tag{4.34}$$

and

$$\langle Q \rangle = \lim_{N\to\infty}\frac{1}{N}\sum_{n=0}^{N-1} Q(x_n) = \int_0^{\alpha} Q(x)f(x)dx = \frac{1}{2}I_{\text{ref}}T_c - \frac{m_2T_c^2}{6} \tag{4.35}$$

Because of $\overline{I}_D = \frac{\overline{Q}}{\overline{T}}, \overline{Q} = \langle Q \rangle, \overline{T} = \langle T \rangle, \overline{I}_D = \frac{\overline{V}_{\text{out}}}{R}$, and $\overline{V}_{\text{out}} = (1+\alpha)V_{\text{in}}$, the reference current I_{ref} can be expressed as

$$I_{\text{ref}} = \frac{(1+\alpha)^2 V_I}{R} + \frac{\alpha V_I T_c}{3L} \tag{4.36}$$

A comparison among I_{ref}s calculated by Equation 4.32, by Equation 4.36, and by experiment is shown in Figure 4.15. It is verified that the estimation of I_{ref} with the invariant probability distribution is much more accurate.

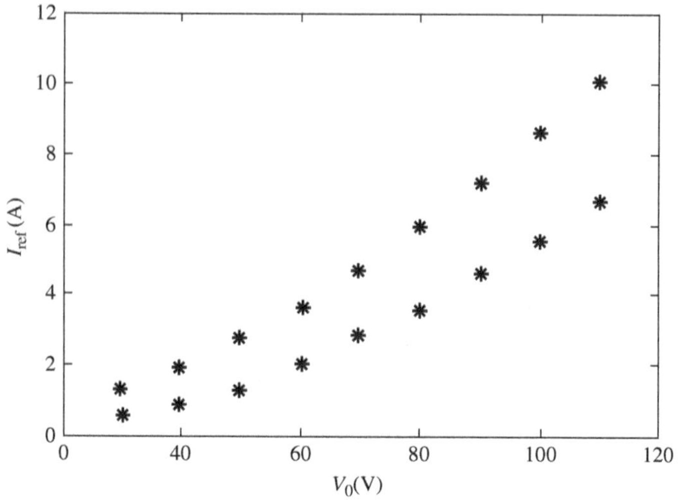

Figure 4.15 Reference current I_{ref} with changing V_{out} obtained by Equation 4.28 ("*"), Equation 4.33 ("x"), and experiments (".")

4.6 Conclusions

The general analytic method for a DC–DC converter was given, and based on these methods, we obtained the discrete-time maps of a DC–DC converter. When the parameters took special values, we got chaotic maps. Because of the inner randomness of chaos, the calculation about chaos must be based on the theory of probability, namely based on the calculation of invariant density distribution. This chapter calculated the invariant density distribution of a chaotic map by using the eigenvector method. And by this result we calculated the power spectral density of the input or output of a DC–DC converter, the average switching frequency, and even estimated the parameter for the circuit. This analytic method is different from the digital method, it can explain the phenomena at root and it can supply the rationale for control EMI in a DC–DC converter to some extent.

References

[1] Hamill, D.C., Deane, J.H.B., Ashwin, P. *et al.* (1999) Calculation of the periodic spectral components in a chaotic dc-dc converter. *IEEE Trans. Circuits Syst.-I*, **46**(11), 1313–1319.

[2] Isabelle, S.H. (1995) A signal processing framework for the analysis and application of chaotic systems. PhD dissertation, Institute of Technology, Cambridge, MA.

[3] Guldner, O., Baranovski, H., Schwarz, A.L. *et al.* (2002) New method for tailoring ripple and spectral properties of chaotic dc-dc converters. *Proc. IEEE Int. Symp. Circuits Syst.*, **4**, 556–559.

[4] Hao, B.L. (1992) *Starting from Parabola*, Science and Technology Publishing Company, Shanghai (in Chinese).

[5] Hao, B.L. (1989) *Elementary Symbolic Dynamics*, World Scientific Press, Singapore.

[6] Dai, X., Zhang, J. and Xu, P. (1999) Calculation of invariant probability distribution for chaotic maps using eigenvector method. *J. Appl. Sci.*, **4**(17), 417–418 (in Chinese).

[7] Li, H., Li, Z., and Halang, W. (2006) Calculation of the invariant density of the chaotic mapping in a DC-DC converter and its applications. 51st Internationales Wissenschaftliches Kolloquium, pp. 1–12.

[8] Woywode, O., Webel, J., Güldner, H. *et al.* (2001) Qualitative dynamics of the boost converter. *IEEE PESC*, 439–444.

[9] Woywode, O., Webel, J., Güldner, H. *et al.* (2003) Bifurcation and statistical analysis of DC-DC converters. *IEEE Trans. Circuits Syst. I*, **50**(8), 1072–108.

[10] Rényi, A. (1957) Representation for real numbers and their ergodic properties. *Acta Inf. Hungarica*, **8**(3–4), 477–493.

[11] Lasota, A. and Mackey, M. (1994) *Chaos, Fractals and Noise: Stochastic Aspects of Dynamics*, Applied Mathematical Sciences Series, vol. **9**, Springer-Verlag, New York.

[12] Eckmann, J.P. and Ruelle, D. (1985) Ergodic theory of chaos and strange attractors. *Rev. Mod. Phys.*, **57**(3), 617–656.

[13] Costa, P.C., Carvalho, A.S., Chouzal, M.F. *et al.* (1998) Direct torque control-frequency analysis. *Adv. Motion Control*, **98**, 21–23.

5

EMI and EMC of Switching Power Converters

5.1 Introduction

Switching power converters have proliferated since the 1970s because of their flexible functions and high power transfer efficiency. Unfortunately, the deterioration of power quality and electromagnetic environment appear synchronously because of the working characteristics of these power electronics equipments. Because of their fast switching speed, switching power converters can generate unwanted electrical signals, which affect other electronic systems. These unwanted signals occur at high frequencies and give rise to electromagnetic interference (EMI). Meanwhile, the low-level gate control circuit of the power converter can also be affected by EMI generated by its own high power circuitry [1]. Moreover, the wideband signals of power electronics equipment can travel over considerable distances and can pollute the electromagnetic environment, because such equipment is usually connected to the utility power line [2]. Consequently, the generated EMI noises need to be controlled and minimized by effective approaches.

Varieties of standards for regulating EMI have become stronger in recent decades. For a good electronic product, the technical level of the circuit design and electromagnetic compatibility (EMC) design, product quality, and technical performance indicators play a crucial role. How to suppress EMI and improve the product quality to satisfy the EMC standards becomes a principal question for engineers of power switching converters.

This chapter reviews the fundamentals of EMI, including the basic terminology and categories of EMI, the various coupling and propagation mechanisms, the causes and effects of low-frequency disturbances, the generation of high-frequency disturbances (both differential mode and common-mode), and the various EMC regulations. A flyback converter is used to illustrate the source and spreading route of EMI of switching converters.

Chaos Analysis and Chaotic EMI Suppression of DC-DC Converters, First Edition. Bo Zhang and Xuemei Wang.

5.2 EMI Origin of Electric Circuits

There are always many resistors, capacitors, inductors, transforms, and switching transistors in electric circuits. The electric field must take place around all electriferous components when the circuit has voltage. Moreover, the magnetic field must take place around all electriferous components when the circuit has current. The capacitor is the component that has the most concentrated electric field between two polar plates. There could be radiated antennae when two plates open wide. Here, any circuit located between the two plates, no matter whether it is open loop or close loop, may generate an induction to the electric field. If a conductor is consistent with the direction of the electric field, it will have a displacement current.

Analogously, the inductor and transformer are components that have the most concentrated magnetic field. The transformer is a typical example of electromagnetic induction, the current of the secondary windings is an inductive current of the primary windings. All the circuits surrounding the transformer can be regarded as secondary coils, and these circuits will have an inducted current when magnetic lines of force or transformer leakage inductance cross it. Two adjacent loops also bring electromagnetic induction, for example, one loop could be regarded as the primary winding of transformers, another loop could be regarded as the secondary winding. Thus, they will disturb each other.

In electronic circuits, EMI will be produced as long as the electric or magnetic field appears. High-frequency signal lines, the pins of the integrated circuits, power switching devices, and various types of connectors are likely to become the source of interference with the radiation of the antenna characteristics, which can emit electromagnetic waves and affect the normal working of other systems or subsystems of its own system. In the case of a PCB (printed circuit board) soldered capacitors and semiconductor devices, the capacitors and soldering act like antennas. They generate and absorb electromagnetic fields. The chips on these boards are so close to each other that the chances of EMI are significant.

5.3 Characteristics of Switching Processes of Power Semiconductors

The voltage conversion is implemented by the high speed switching actions of switching power electronics in the power electronic devices. The key components of power electronics devices are power semiconductors, such as the power diode, metal oxide semiconductor field effect transistor (MOSFET), insulated gate bipolar transistor (IGBT), Silicon Controlled Rectified (SCR), Gate Turn-Off Thyristor (GTO), and so on. Regardless of the main circuit or control circuits during the switching processes of power semiconductors, there are always high dv/dt and di/dt ratings that can cause transient electromagnetic noise. The frequencies of such noise can be up to tens of kilohertz, to a few hundred kilohertz or even several megahertz, so it cannot be ignored.

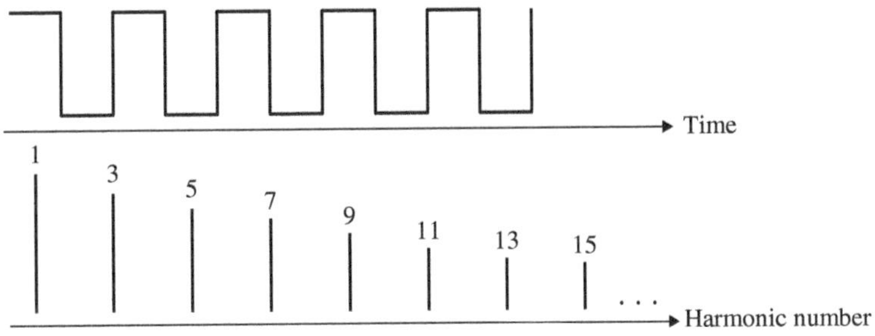

Figure 5.1 Square waveform (up) and its spectrum (down)

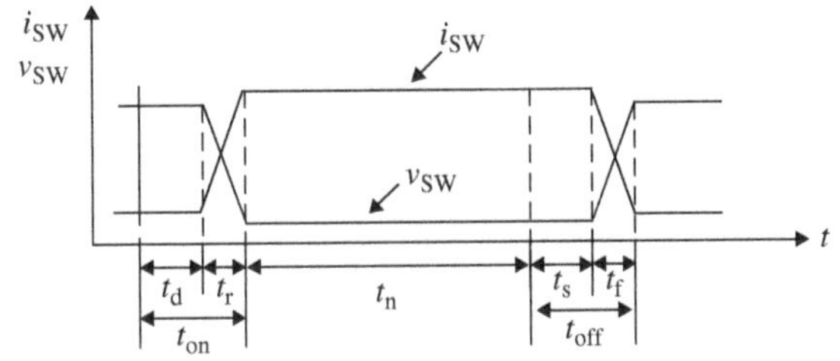

Figure 5.2 Topical waveforms of device voltages and currents

Current and voltage waveforms of ideal switches in power electronics converters are square waves. The frequency spectrum calculated by the Fourier transform of a square wave consists of a fundamental at the square wave frequency, plus the odd harmonics decreasing in size shown in Figure 5.1.

However, the characteristics of a practical switching transistor differ from those of an ideal switch. During the turn-on and off process, a practical switching device as shown in Figure 5.2, requires a finite delay time t_d, rise time t_r, storage time t_s, and fall time t_f. As the device current i_{SW} rises during turn-on, the voltage across the device v_{SW} falls. As the device current falls during turn-off, the voltage across the device rises [1]. The typical waveforms of device voltages v_{SW} and current i_{SW} are shown in Figure 5.2.

The most significant spectrum envelope of practical switching waveforms is shown in Figure 5.3. There are two inflections on the envelope, one is at $1/\pi t_f$, and the other is at $1/\pi t_r$. When the frequency is less than $1/\pi t_f$, the amplitude spectral envelope is fixed and equals $20\lg(2At_fT)$. When the frequency is between $1/\pi t_f$ and $1/\pi t_r$, the amplitude spectral envelope is the 20 dB/decade drop. And when the frequency is larger than $1/\pi t_r$, the spectral envelope decreases is 40 dB/decade drop. Obviously, once t_r or t_f increase, the amplitude of harmonic spectrum will decrease.

Here, a MOSFET will be taken as an example to analyze the EMI caused by switching behavior. A MOSFET is a commonly used power semiconductor, and its equipment circuit diagram with parasitic elements is shown in Figure 5.4, where C_{gs}, C_{gd},

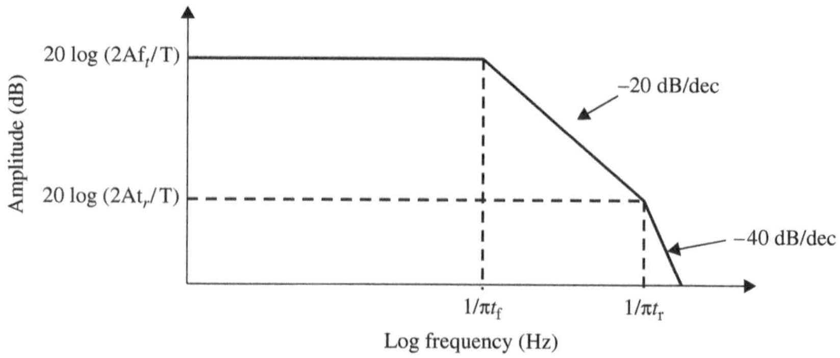

Figure 5.3 Amplitude spectral envelopes of periodic waves

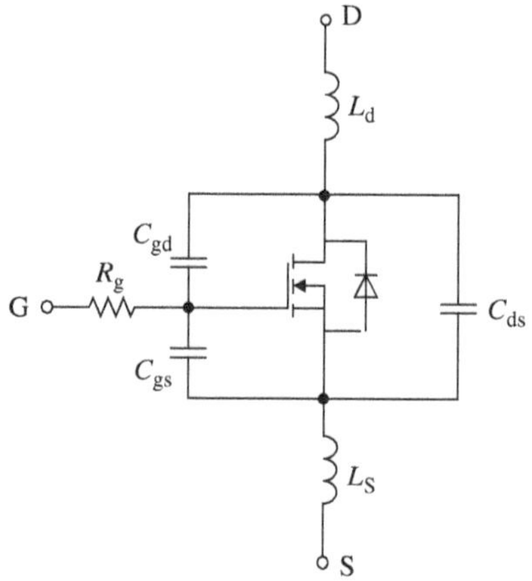

Figure 5.4 Simplified equipment circuit diagram of MOSFET with parasitic elements

and C_{ds} are parasitic capacitances of the PN junction, R_g is the internal gate resistance, and L_s, and L_d are the parasitic connection inductances presented in the components and connections respectively. These parasitic parameters are determined by the structural, internal capacitances, the internal and terminal resistances of the MOSFET. At the turn-on instant, parasitic capacitors discharge quickly and high-current spikes caused by the high rate of change of voltage (*dv/dt ratings*) flow through the MOSFET. At the turn-off instant, the energy stored in the parasitic inductance may discharge to the parasitic capacitor, so high-voltage spikes caused by the high rate of change of current (*di/dt ratings*) produce and trigger high-frequency parasitic oscillations, also called ringing, shown in Figure 5.5.

The rise time and fall time of the current and voltage of the switching transistor are always between 10 and 100 ns, so the slew rate *di/dt* and *dv/dt* are very fast. According to the spectrum analysis, a higher slew rate will cause a higher amplitude harmonic.

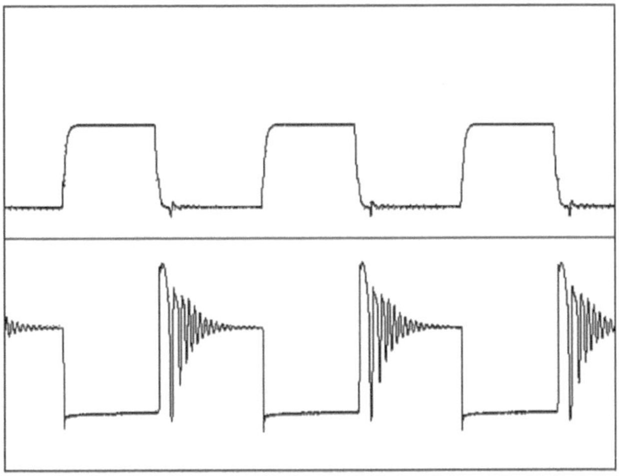

Figure 5.5 Ringing voltage waveforms caused by parasitic elements. The up waveforms are the drive signal of MOSFET (5 V/div, 10 μs/div), the below waveforms are v_{DS} (50 V/div, 10 μs/div)

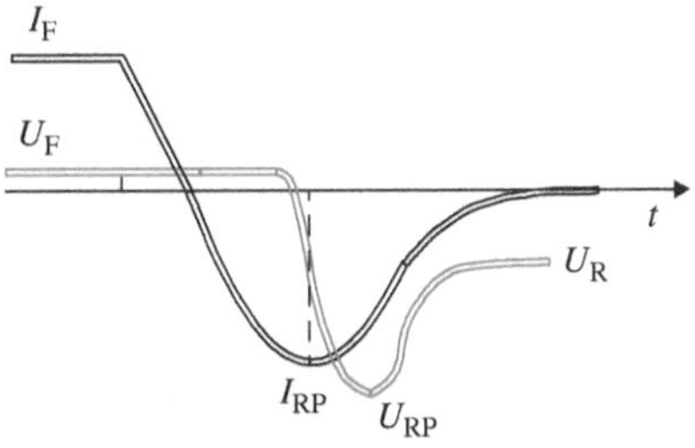

Figure 5.6 Turn-off behavior of a diode

These wideband harmonics will further cause strong interference voltage or current, and produce EMI by some coupling approaches. Moreover, the power electronics equipment is usually connected to the utility line, such that the wideband harmonics are able to travel over considerable distances and pollute the electromagnetic environment with unwanted interference [2].

Another commonly used component in DC–DC converters is an uncontrolled device, the power diode. In practice, the diode has a special reverse recovery characteristic as shown in Figure 5.6. When a diode is positive based, a current flows through it and it is in conductive state. But at the instant of turn-off, the diode has a quite large reverse current accompanying obvious reverse overvoltage, and the diode loses its reverse blocking capability during this brief period of time. Due to the large amplitudes of the reverse recovery current and *di*/*dt*, a high induction voltage will be produced with the lead inductance and other connected circuits, resulting in strong transient EMI.

5.4 Overview of EMI and EMC

5.4.1 Basic Principles of EMI

There are a number of key terms used in EMI and EMC. It is worth spending some time developing an understanding of the basic principles and properties of EMI. Most of the terms are taken from [2–4].

Electromagnetic interference (EMI): The degradation in performance of a device, equipment, or system caused by electromagnetic disturbance.

Electromagnetic compatibility (EMC): (i) The ability of a device, equipment, or system to function satisfactorily in its electromagnetic environment (immunity aspect) without introducing intolerable electromagnetic disturbances to anything in that environment (emission aspect). (ii) An engineering discipline developed to ensure that devices, equipment, or systems that generate and/or use electromagnetic energy can coexist satisfactorily [2, 3].

EMI is categorized as **conducted EMI** or **radiated EMI** in terms of frequency. Electromagnetic radiation and electromagnetic conduction are differentiated by the way an electromagnetic field propagates. Conducted EMI is caused by the physical contact of the conductors as opposed to radiated EMI which is caused by induction (without physical contact of the conductors). EMI can be conductive and/or radiative. Its behavior is dependent on the frequency of operation and cannot be controlled at higher frequencies. For lower and higher frequencies, EMI is caused by conduction and radiation respectively. For example, the skin effect is due to conducted EMI and the proximity effect is due to radiated EMI. Conducted EMI normally has a frequency between 150 kHz and 30 MHz. The frequency band of radiated EMI is from 30 MHz to 1 GHz. In practical terms, the conducted interference should be suppressed first, thus about 80% of radiated interference may be eliminated.

The schematic descriptions can be simply illustrated in Figure 5.7. The basic arrangement of noise source, coupling path and victim, receptor or sink is shown in Figure 5.7. The source and victim are usually electronic hardware devices, though the source may be a natural phenomenon such as a lightning strike, electrostatic discharge (ESD) or, in one famous case, the Big Bang at the origin of the Universe. There are four fundamental coupling mechanisms of EMI, including galvanic, inductive, capacitive, and radiated. Any coupling path can be broken down into one or more of these coupling mechanisms working together. For example the lower path in the diagram involves inductive, conductive, and capacitive modes.

Galvanic coupling: The disturbing current flows in a common circuit impedance. This coupling can distribute through interface cables, antenna input terminals, and metallic shells.

Radiated coupling: When the interferences pass through free space, or some other nonconductive medium, this coupling is radiated coupling. This kind of coupling

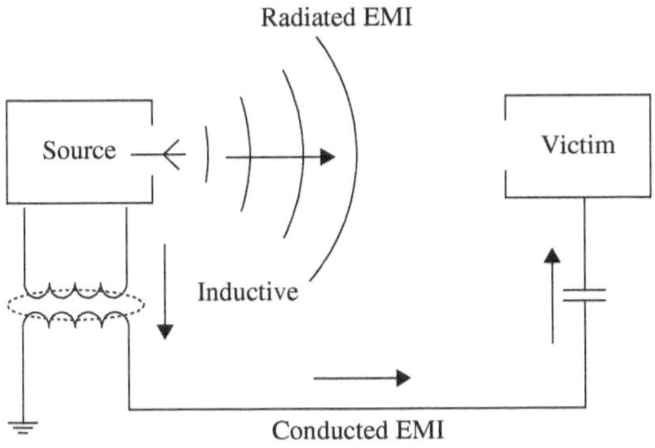

Figure 5.7 Four electromagnetic interference coupling modes [5]

occurs if the distance between the source device and the receptor is several wavelengths and distributes by electromagnetic wave.

Inductive coupling or *magnetic coupling*: This occurs when a varying magnetic field exists between two parallel conductors, inducing a change in voltage along the receiving conductor.

Capacitive coupling: This occurs when a varying electrical field exists between two adjacent conductors typically less than a wavelength apart, inducing a change in voltage across the gap.

Inductive coupling and capacitive coupling are all couplings through the electromagnetic near field (typically less than a wavelength apart), only, dominated by the magnetic and electric field respectively. So, generally speaking, inductive coupling and capacitive coupling are often collectively referred to as inductive coupling.

The conducted EMI can be further classified into differential-mode (or symmetrical) and common-mode (or asymmetrical). Differential-mode propagation takes place between two conductors, as phase line and neutral line, which form a conventional return circuit. On the other hand, common-mode propagation takes place between a group of conductors and the ground (Figure 5.8). Even if the ground connection is not deliberate, common-mode currents flow through parasitical capacitors or parasitical inductors to the ground.

5.4.2 *EMC Regulations*

The EMC of electronic and electrical equipment has been paid a great deal of attention as a growing problem and has resulted in directives to the manufacturers which set out the essential manufacturer requirements before their equipment can be marketed or sold. Organizations in individual nations, across Europe and worldwide, were set up to maintain these directives and the associated standards. This regulatory environment

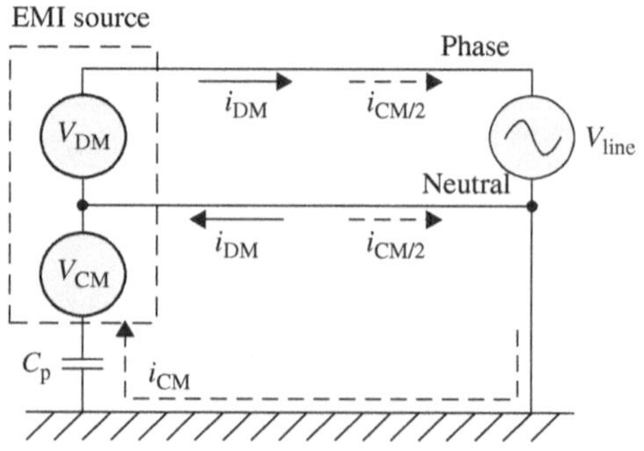

Figure 5.8 Differential-mode and common-mode conducted EMI [6]. © [1996] IEEE. Reprinted, with permission, from Kodali, V.P., Engineering Electromagnetic Compatibility: Principles, Measurements, and Technologies, IEEE Press, NY, 1996

led to a sharp growth in the EMC industry supplying specialist devices and equipment, analysis and design software, and testing and certification services.

Regulatory work to ensure interference-free reception started in 1933 with the formation of the International Special Committee on Radio Interference (CISPR) [7]. CISPR's standards cover the measurement of radiated and conducted interference, and set various standards for the test layout, to help improve the reliability of comparison between tests. EMI test results can vary widely according to the exact layout of the equipment and cabling. These standards cover cable lengths, measurement device configuration, and grounding schemes. The standards also address immunity from external interference. The needs of commercial and military equipment are different. The commercial standards specify the requirements to protect radio, telecommunication, television, domestic, and industrial systems [1]. There is a bewildering variety of national and international EMC standards, regulating conducted and radiated emission and susceptibility of equipment and systems. These regulations are mostly based on EMC standards issued by CISPR, for example, the EN standards in European Union, the BSI in the United Kingdom, and the VDE in Germany, and so on, as shown in Table 5.1. These standards attempt to standardize product EMC performance, with respect to radio interference for electrical equipment.

Table 5.1 EMC standards

EMC standards	
United States FCC	FCC PART15
European Union EN	EN55022/11
United Kingdom BS	BS4727
Germany VDE	VDE0228
Japan VCCI	CISPR22
China CCC	GB9254

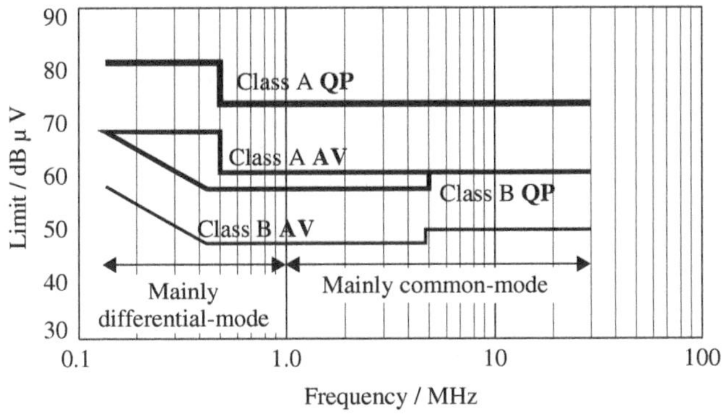

Figure 5.9 Limit of conducted interference of CISPR 22

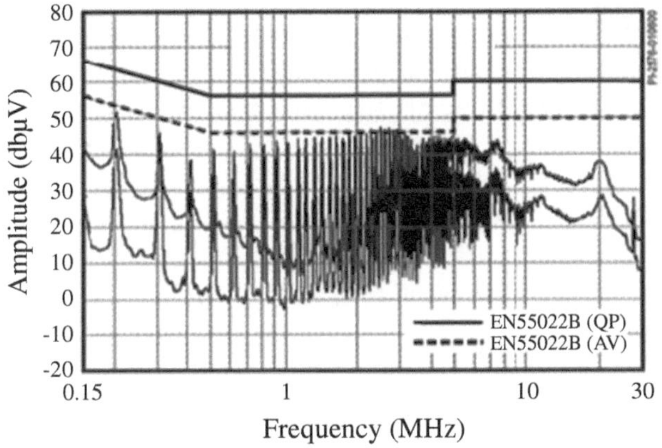

Figure 5.10 Full range conducted EMI scan of a typical flyback converter using Topswitch-II (100 kHz)

Figure 5.9 shows the widely used CISPR 22 conducted noise limits. CISPR 22 classifies approvals into two classes, Class A and Class B, as shown in Figure 5.9. Class A applies to the industrial or business environment, and Class B applies to the residential environment. Since Class B devices are more likely to be located in closer proximity to radio and TV receivers, the emission limits for Class B are about 10 dB more restrictive than Class A, that is, Limit class A < Limit Class B. Each class has two levels, quasi-peak and average value. Figure 5.10 is an actual full range conducted EMI scan of a typical flyback converter using TOP220 as switching transistor.

5.5 EMI of Power Electronic Converters

Here, an example is utilized to analyze the source and route of electromagnetic disturbance of power electronic converters. A typical flyback converter topology

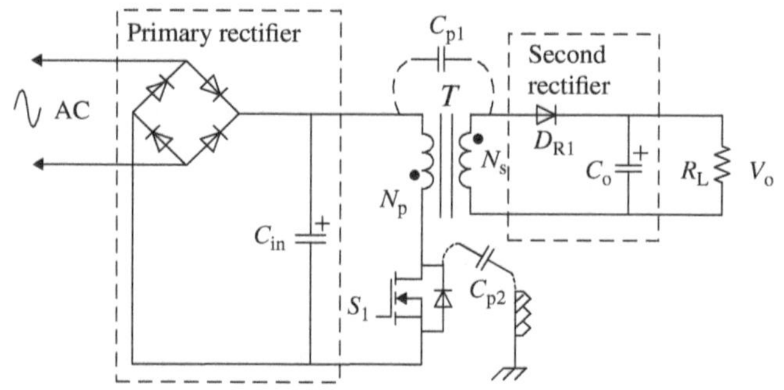

Figure 5.11 A typical flyback converter [3]. © [1996] IEEE. Reprinted, with permission, from Redl, R., ELFI, S. A., Onnens, Power electronics and electromagnetic compatibility. Proceeding of Power Electronics Specialists Conference, 1996

is shown in Figure 5.11. Since flyback converters need very few components, it is a very popular topology for low- and medium-power applications such as battery chargers, adapters, and DVD players [3]. This converter consists of a primary rectifier, transformer T, switching transistor switch S_1 and secondary rectifier, parasitic capacitance C_{p1} and C_{p2}.

5.5.1 Parasitic Parameters of Flyback Converters

5.5.1.1 Switching Transistors

The heatsink must always be affixed in the switching transistors to be radiating because of their high working current. For reasons of safety and preventing the radiation of high frequency EMI, the heatsink is always connected to the protective ground and equipment's shell. Meanwhile, a heat conductive insulating shim is pasted between the heatsink and the switching transistor to prevent short-circuit faults. Thus, the parasitic capacitor, which cannot be ignored in high frequency working, is formed (Figure 5.12). The high frequency switching current will pass

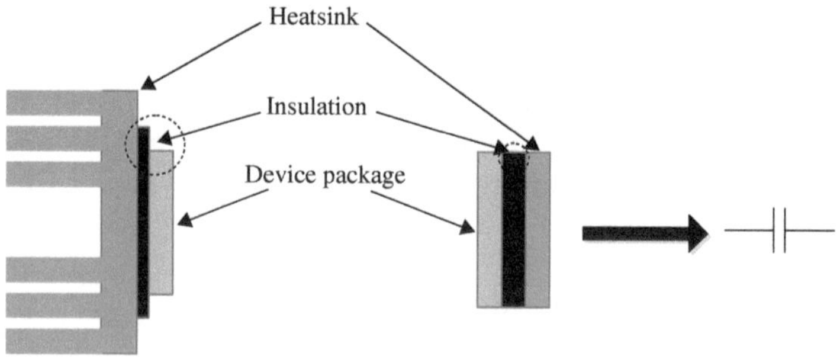

Figure 5.12 Parasitic coupling capacitance of the device package and the heatsink

through this parasitic capacitor to the heatsink first. Then, it flows to equipment shell and protective ground. Finally, the radiated interference to space and common-mode to the power line propagate constantly.

Supposing the voltage of the switching transistor is about 300 V, and the falling and rising time of voltage waveforms reaches 100 ns, so the slew rate is 300 V/100 ns or 3 kV/1 μs. If the parasitic capacitor between switch baseplate and grounded heatsink is 50 pF, the instant common-mode current to ground reaches

$$i = C\frac{du}{dt} = 50 \times 10^{-12} \times (3000/10^{-6}) = 150\ \text{mA}$$

5.5.1.2 High Frequency Transform

1. *Magnetic inductance.* Magnetic inductance is the primary inductance of a pulse transformer. Actually, it is an equivalent inductance and it is used to generate the magnetism of the iron core, so that the molecule of ferromagnetic can be magnetic. The iron core is neutral originally, but it is like a permanent with winding the wire and adding the power.
2. *Leakage inductance.* Leakage inductance is the property of an electrical transformer that causes a winding to appear to have some inductance in series with the mutually-coupled transformer windings. This is due to imperfect coupling of the windings and creation of leakage flux which does not link with all the turns of the winding. The leakage flux alternately stores and discharges magnetic energy with each electrical cycle, thus it effectively acts as an inductor in series in each of the primary and secondary circuits. Leakage inductance is primarily caused by the design of the core and the windings. Voltage is dropped across the leakage reactance, resulting in poorer supply regulation when the transformer is placed under load.
3. *Distributed capacitance.* There are distributed capacitances in the actual winding of the transformer, especially, existing between the coil wire, between the transformer cores, and between windings, as shown in Figure 5.13. The amount of the capacitance depends on the winding geometry, the dielectric constant of the

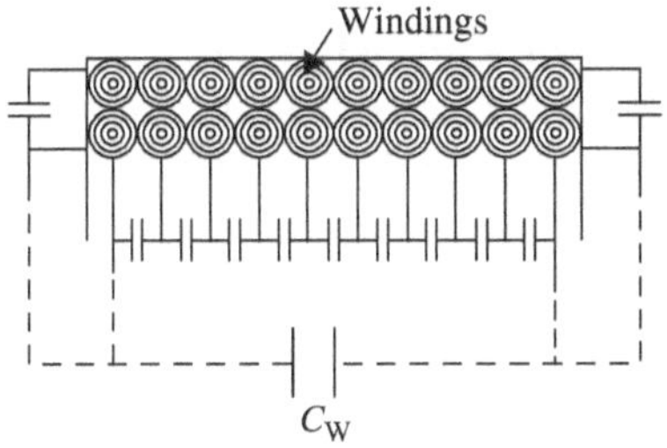

Figure 5.13 Parasitic capacitances of a transform

core material and its packaging materials, and so on. The voltage of distributed capacitance changes only slowly in low-frequency circuits, so the extra current is usually negligible. In high-frequency circuits, however, the voltage changes quickly. Thus, the extra current caused by the distributed capacitance is larger and cannot be negligible.

There are many parasitic capacitors in this converter, but only the two main capacitors C_{p1} and C_{p2} are used to analyze the EMI spreading routes for reasons of simplification. C_{p2} is parasitic capacitance of the switching component S_1 and heatsink, and C_{p1} is parasitic capacitance of the primary and secondary windings of the transformer.

5.5.2 Primary Rectifying Circuit

The rectifying circuit is a common factor causing EMI because it is connected to the AC power grid directly. The sine line voltage becomes a single phase pulse voltage after rectifying. And the high order harmonic of this pulse current and voltage will be introduced to the power line coupled by conducted interference, and would interfere with the electrical equipments connected to the same power grid.

The primary rectifier of the flyback converter is a typical single-phase rectifier consisting of the four-diode bridge rectifier with capacitive filter C_{in}. Figure 5.14 shows the line current and voltage waveforms of the primary rectifier, where the current waveforms are a series of narrow distortion pulses. By spectrum analysis in FFT (Fast Fourier Transform), the current waveforms have plenty of high order harmonics and Total Harmonic Distortion (THD) reaches 109.62%. This distorted line-current makes the quality of the line power decline, causes deteriorating power factor, and reduces the maximum power available from a line power. Meanwhile, the harmonic currents also have a number of undesirable effects, including excess audio noise, overheating of transformers, generators, motors, and mechanical oscillations in generators and motors [1].

5.5.3 Switching Loop

The switching loop, a key component of the switching converter, consists of a switching transistor and a high frequency transformer. Figure 5.15 shows several parasitic components of a switching loop with magnetizing inductance L_m, such as primary and secondary leakage inductors L_{lk1}, an output capacitor of MOSFET C_{oss}, and a junction capacitor of a secondary diode C_j. The flyback converter may operate in both the CCM (continuous conduction mode) and DCM (discontinuous conduction mode). The voltage and current waveforms of the switching transistor are shown in Figure 5.16.

The primary current i_d alters the capacitor of the MOSFET C_{oss} in a short time when the MOSFET is turned off. Moreover, when C_{oss} exceeds the input voltage plus reflected output voltage $V_{in} + nV_{out}$, the secondary diode is turned on. And then the

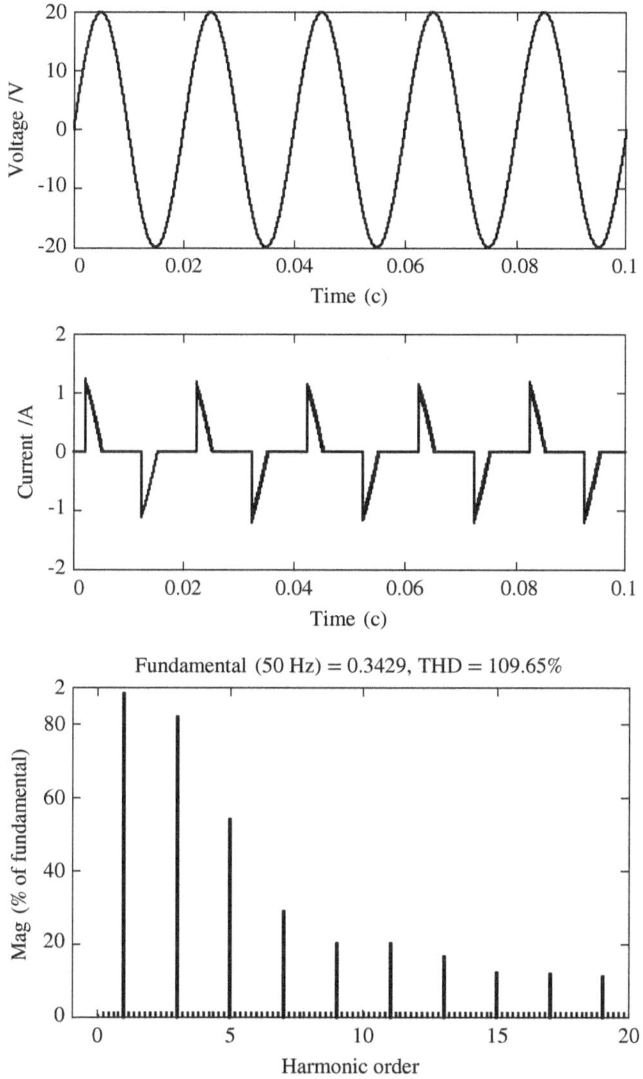

Figure 5.14 Current and voltage waveforms of a primary rectifier [1]

voltage across the magnetizing inductor L_m clamps to nV_{out}. Therefore, a resonance occurs between L_{lk1} and C_{OSS} with a high-frequency and high-voltage surge, which will cause failure [6].

The reverse recovery current of the secondary diode increases the primary current when the MOSFET is turned on. So the primary current has a large current surge at the instant of turn-on. When the secondary current runs dry at the end of the switching period in DCM, another resonance takes place between L_m and C_{OSS} of the MOSFET [6].

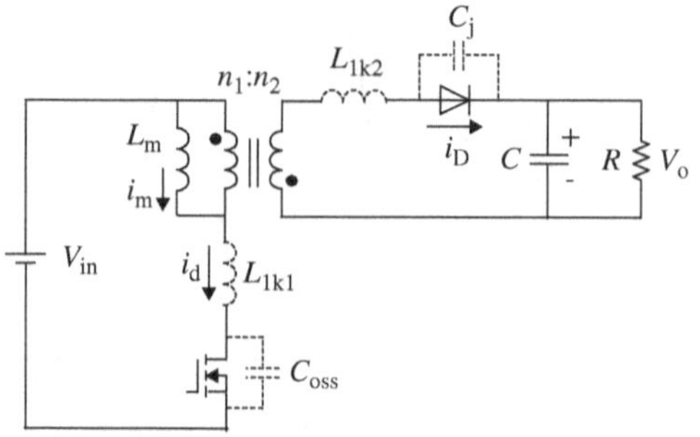

Figure 5.15 Configuration with parasitic components of a basic flyback converter [3]

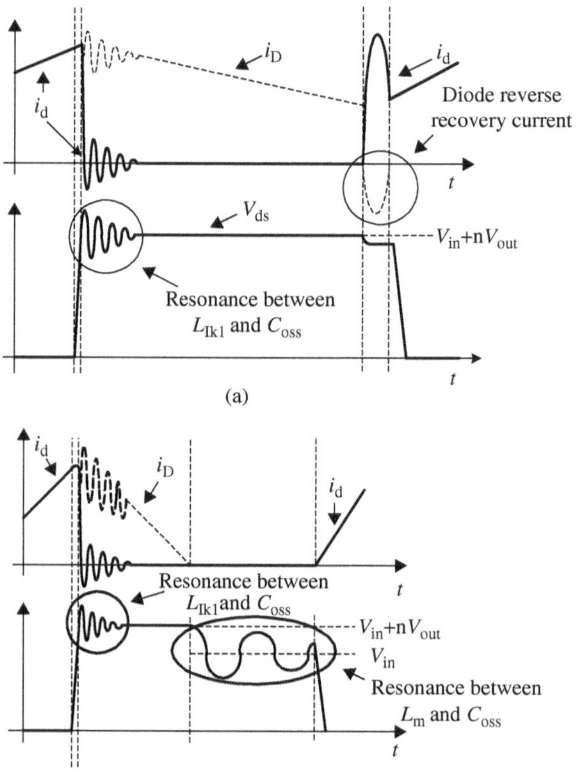

Figure 5.16 (a) CCM operation and (b) DCM operation of a flyback converter [3]

Thus, strong different-mode interferences, which have plenty of high order harmonics whose frequency can reach to hundreds of megahertz, are caused by these high overvoltages and oscillations. This interference can also change to common mode and radiated interference through the inductive coupling in the inner circuit of the

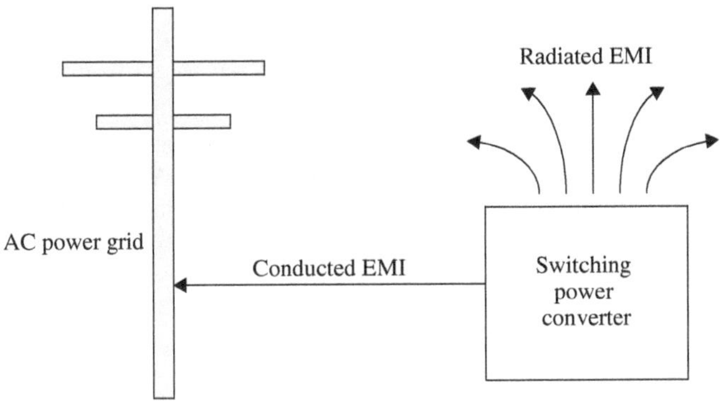

Figure 5.17 EMI of a switching power converter

converters, where the intensity of radiated EMI is in direct proportion to the area of the circuit board, the current, and the frequency of ringing.

5.6 Conclusions

The EMI of switching power converters is generated inside the power supply devices; conducted and radiated interference spread to the AC power grid (shown in Figure 5.17), power supply units, and space through the high frequency transform, energy storage inductor, parasitic capacitor, and inappropriate system configuration and component placement. There are four characteristics.

1. The EMI of switching power converter is caused by the switching actions of power semiconductor. The source of the EMI mainly focuses on the switching transistor, rectifying diode, and heatsink and high frequency transform connected to them. The EMI originating from the switching loop is the most straightforward and principal interference.
2. The EMI of power electronic equipment has characteristics of wide frequency bandwidths and great disturbance intensity because of high dv/dt and di/dt rates.
3. The main EMI is conducted interference because the switching frequency is not high enough (from several kilohertz to megahertz).
4. Unsuitable wiring of PCB is also a main factor of EMI.

References

[1] Rashid, M.H. (2004) *Power Electronics: Circuits, Devices, and Application*, 3rd edn, Pearson Education.

[2] Redl, R. and Elfi, S.A., Onnens (1996) Power electronics and electromagnetic compatibility. 27th Annual IEEE Power Electronics Specialists Conference, 1996. PESC'96 Record, Baveno, Italy, Vol. **1**, pp. 15–21.

[3] Kodali, V.P. (1996) *Engineering Electromagnetic Compatibility: Principles, Measurements, and Technologies*, IEEE Press, New York, pp. 345–348.
[4] Wikipedia. Electromagnetic Compatibility, http://en.wikipedia.org/wiki/Electromagnetic_compatibility (accessed 7 June 2014).
[5] Li, H., Li, Z., Zhang, B. *et al.* (2009) Suppressing Electromagnetic Interference in Direct Current Converters. *IEEE Circuits And Systems Magazine*, FOURTH QUARTER, pp. 10–28.
[6] FAIRCHILD (2006) Application Note AN-4147 Design Guidelines for RCD Snubber of Flyback Converters, www.fairchildsemi.com. (accessed 7 June 2014).
[7] Redl, R. (2001) Electromagnetic environmental impact of power electronics equipment. *Proce. IEEE*, **89**(6), 926–938.

6

Discrete Subsystem Chaotic Point Process of DC–DC Converters and EMI Suppression

6.1 Introduction

The point process [1, 2] means that its process characteristics are decided by the event time. This physical phenomenon can be represented by the highly localized events distributed randomly in the continuous system. Moreover, each localized event in the model is expressed in terms of an idealized point, and this point is considered in the same location with the event in a continuous system. For example, the time point of a gamma photon produced by a radioactive source in radioactive decay and the special point traffic flow on an expressway can both be described by the point process. Two-dimensional events can also be analyzed by point process, such as the time and location sets about an earthquake, detection of blood flow in nuclear medicine, the dynamic parameters of physiological phenomenon in metabolic exchange, and so on.

The research shows that the power electronic converter is a typical hybrid dynamic system [3]. And it consists of two parts, one is the continuous subsystem described by behavior and the other is the discrete subsystem described by the switching process. The behaviors of the discrete subsystem are decided by the pulse width modulation (PWM) controlled strategy of converter. The discrete subsystem after chaotization can be partly regarded as a point process constituted by PWM pulse signals, the time point characteristics of the switching value can be described and analyzed by chaotic point process theory. In this chapter the spectrum of a converter and its statistical regularity are obtained, and the mechanism of converter chaos suppressing EMI (electromagnetic interference) is explored in depth. This chapter uses an AC–DC converter as an example to analyze the spectral characteristics of fixed frequency PWM, periodical PWM, and chaotic PWM.

Chaos Analysis and Chaotic EMI Suppression of DC-DC Converters, First Edition. Bo Zhang and Xuemei Wang.

6.2 Description of Chaotic Point Process of DC–DC Converters

6.2.1 Model of Chaotic Point Process of DC–DC Converters

The power electronics based on hybrid system theory can be described by a continuous-discrete subsystem, as shown in Figure 6.1 [4]. This system can be expressed by the following equations

$$\frac{dx}{dt} = \phi(x, \mu) \quad x \in X \subset R^n \; t \subset R \; \mu \in R^l \tag{6.1}$$

$$\mu_{i+1} = G(\mu_i)|H(x, t) = 0 \qquad i = 0, 1, 2 \cdots \tag{6.2}$$

Equation 6.1 belongs to a continuous subsystem, and Equation 6.2 belongs to a discrete subsystem. The characters ϕ and G are nonlinear vectors, x is the continuous-time state, and μ is the discrete input or controls parameter. The two subsystems are under the same feedback controlled loop, and have the following characteristics.

1. When the condition $H(x,t)$ is satisfied, the discrete subsystem will finish an iteration, and the μ_k and PWM pulse interval T_k can be calculated.
2. The states of the discrete subsystem as the input will be fed back to the continuous subsystem.

It can clearly be seen that the PWM pulse interval T_k can reflect the duty circle and frequency change of switching variation. When the PWM frequency is fixed, T_k is decided by the feedback current voltage and sawtooth wave. While it is chaotic PWM, T_k will be decided by the feedback current voltage and sawtooth wave whose

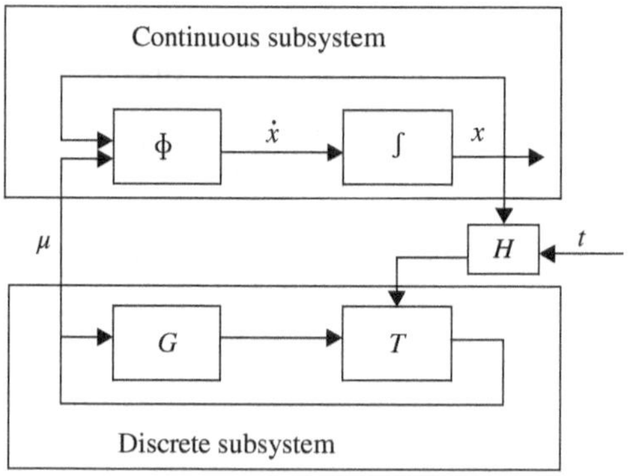

Figure 6.1 Structure of a power converter as a continuous-discrete system [5]. © [2003] IEEE. Reprinted, with permission, from Baranovski, A. L., Schwarz, W., Chaotic and random point processes: analysis, design, and applications to switching systems. *IEEE Trans. Circuits and Systems I: Fundamental Theory and Applications*, 2003

frequency changes in chaos. The different production and means of control of the sawtooth wave will make the PWM drive pulse work in a fixed-frequency or chaotic state.

Whether it is fixed-frequency PWM or chaotic PWM, their PWM pulse frequencies can all be expressed by the following equation

$$f = f_s + \Delta f = v_s + kv_n \qquad v_n = \varphi(v_{n-1}) \tag{6.3}$$

where v_s is the fixed voltage and it is producing fixed nominal switching frequency f_s. v_n is the spread frequency signal voltage generated by chaotic mapping φ, which will decide chaotic switching frequency deviation Δf. And k is the chaotic modulation degree.

It can be seen from Equation 6.3 that the frequency or interval T_k and arrival time τ_k of the chaotic PWM drive pulse signal are all chaotic variations decided by $\varphi(v_f, t)$. The point process of chaotic PWM decided by the discrete subsystem is shown in Figure 6.2, and it can be expressed by the following formula

$$\xi(t) = \sum_{k=1}^{\infty} A\delta(t - \tau_{k-1}) \qquad \Omega = \{\tau_k : k \geq 0\} \tag{6.4}$$

where τ_{k-1} is the arrival time point of the driving pulse, A is the driving pulse amplitude, and Ω is the chaotic point process sets of the discrete subsystem of the chaotic PWM converter. The spectrum character of the chaotic PWM converter is decided by the driving pulse, namely, the chaotic point process.

6.2.2 *Statistical Characteristics of the Chaotic Point Process in Converter*

The interval T_k of the periodical and chaotic PWM pulse presents periodic variation and chaotic variation respectively [6]. The starting time τ_k of number k driving wave is the accumulation of the interval T_k, and it meets

$$\tau_k = \tau_{k-1} + T_{K-1} \tag{6.5}$$

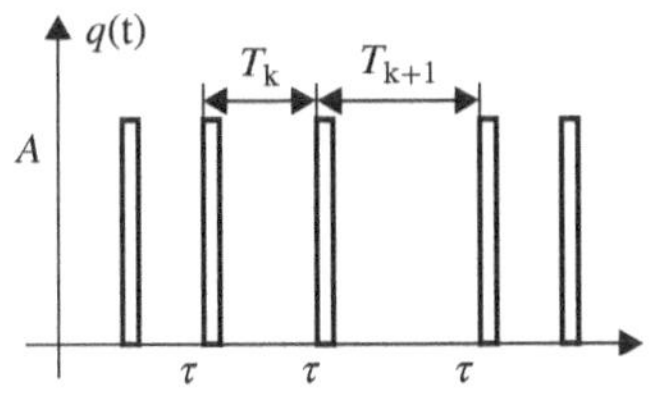

Figure 6.2 Chaotic point process of discrete subsystems in power converters

If $\tau_0 = 0$, it can obtain $\tau_k = \sum_{i=1}^{k} T_i$, so the starting time τ_k of the number k PWM pulse represents periodicity and chaos with different spread frequency signals.

Thus, the joint probability density function of sequence τ_k is

$$f_{\tau_1,\cdots\tau_k}(x_1,\cdots,x_k) = \frac{\partial^k F(\tau_1 < x_1,\cdots \tau_k < x_k)}{\partial x_1 \cdots \partial x_k} = f_\tau(x_1)\prod_{i=1}^{k-1}\delta(x_{i+1} - \phi^{(i+1)}(x_1)) \quad (6.6)$$

The characteristic function about τ_k is

$$\begin{aligned}\Theta_{\tau_1,\cdots,\tau_K}(\omega_1,\cdots,\omega_k) &= E\left(\exp\left(j\sum_{i=1}^{k}\omega_i\cdot\tau_i\right)\right)\\ &= \int_x\cdots\int\left(\exp\left(j\sum_{i=1}^{k}\omega_i\cdot x_i\right)\right)f_{\tau_1,\cdots,\tau_K}(x_1,\cdots,x_k)\end{aligned} \quad (6.7)$$

According to sequence τ_k $(k=0,\ 1,\ \ldots)$, a continuous-time process $N(t)$ can be defined as follows

$$N(t) = \max\{k : \tau_k \le t\} \quad (6.8)$$

where $N(t)$ represents the PWM pulse number during interval $[0,\ t)$ it is called the counting process or renewal process of the PWM driving pulse process $\xi(t)$. The counting process $N(t)$ of the PWM driving pulse process $\xi(t)$ is shown in Figure 6.3.

The discrete probability distribution $F_k(T)$ of $N(t)$ is

$$F_k(T) = P(N(t) = k) = F_\tau(t,k) - F_\tau(t,k+1) \quad (6.9)$$

$F_\tau(t,\ k)$ in Equation 6.9 is the arrival time distribution of the number k chaotic PWM drive pulse.

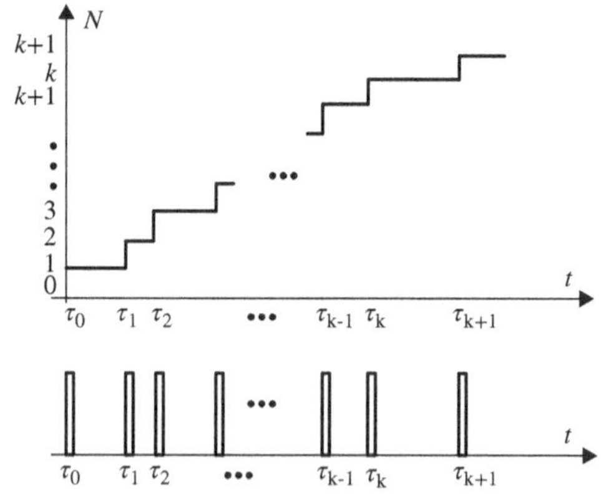

Figure 6.3 Counting process $N(t)$ of the PWM drive pulse

6.3 Spectral Quantification Analysis of the PWM Pulse Process

6.3.1 Spectral Quantification Analysis of the Periodic PWM Pulse

In Equation 6.3, let $\Delta f = f_A \sin(2\pi f_m t)$ be the sinusoid periodic spread spectrum signal, then the PWM switching frequency is [4, 6]

$$f = f_s + f_A \sin(2\pi f_m t) \tag{6.10}$$

where $f_m = 1/T_m$ is the frequency of the spread sinusoid signal, f_A is the maximum frequency deviation of the spread spectrum PWM driving pulse.

If $\Delta f = \begin{cases} f_A & nT_m \le t < nT_m + T_m/2 \\ 0 & nT_m + T_m/2 \le t < (n+1)T_m \end{cases}$, the periodic spread spectrum signal is square wave, so the PWM switching frequency is

$$f = \begin{cases} f_s + f_A & nT_m \le t < nT_m + T_m/2 \\ f_s & nT_m + T_m/2 \le t < (n+1)T_m \end{cases} \tag{6.11}$$

If $\Delta f = \frac{f_A(t\mathrm{mod}T_m)}{T_m}$, the periodic spread spectrum signal is a sawtooth, so the PWM switching frequency is

$$f = f_s + \frac{f_A(t\mathrm{mod}T_m)}{T_m} \tag{6.12}$$

The instantaneous frequency nth degree harmonic with the PWM driving pulse of periodical spread frequency is

$$f_{\mathrm{inst},n} = \frac{1}{2\pi}\frac{d\theta_n}{dt} = n[f_s + f_A \sin(2\pi f_m t)] \tag{6.13}$$

where θ_n is the phrase of the nth degree harmonic, thus

$$\theta_n = 2\pi n \int_0^t [f_s + f_A \sin(2\pi f_m t)]dt = 2\pi n f_s t - n\beta \cos(2\pi f_m t) + n\beta \tag{6.14}$$

where, $\beta = \frac{f_A}{f_m}$ is defined as the parameter of the periodic spread spectrum.

Using a Fourier sequence to describe the PWM pulse process $\xi(t)$, and substituting θ_n into $\xi(t)$, one can obtain

$$\begin{aligned} \xi(t) &= \sum_{n\to-\infty}^{\infty} C_n e^{j\theta_n} \\ &= \sum_{n\to-\infty}^{\infty} C_n e^{j(2\pi n f_s t + n\beta)} \left\{\cos\left[n\beta\cos(2\pi f_m t)\right] - j\sin\left[n\beta\cos(2\pi f_m t)\right]\right\} \end{aligned} \tag{6.15}$$

Using the Jacobi equation [7], cos and sin can be expressed as

$$\cos[n\beta \sin(2\pi f_m t)] = J_0(n\beta) + \sum_{k=even}^{\infty} 2J_k(n\beta)\cos 2\pi k f_m t$$

and

$$\sin[n\beta \sin(2\pi f_m t)] = \sum_{k=odd}^{\infty} 2J_k(n\beta)\sin(2\pi k f_m t)$$

respectively.

Therefore, we get

$$G(t) = \sum_{n\to-\infty}^{\infty} C_n \left\{ e^{j2\pi n f_s t} \cdot J_0(n\beta) + \sum_{k=1}^{\infty} J_k(n\beta) \times \left[e^{j2\pi t(kf_m+nf_s)} + (-1)^k \sum_{k=odd}^{\infty} J_k(n\beta) e^{j2\pi t(nf_s-kf_m)} \right] \right\} \tag{6.16}$$

Here $J_k(\cdot)$ is a k-order Bessel function, and it can be expressed by the following sequence

$$J_k(n\beta) = \left(\frac{n\beta}{2}\right)^k \left[\frac{1}{k!} - \frac{\left(\frac{n\beta}{2}\right)^2}{1!(k+1)!} + \frac{\left(\frac{n\beta}{2}\right)^4}{2!(k+2)!} - \frac{\left(\frac{n\beta}{2}\right)^6}{3!(k+3)!} + \cdots \right] \tag{6.17}$$

It has two characters as follows

$$J_k(n\beta) \approx 0, k > n\beta + 1 \tag{6.18}$$

$$\sum_{k=-\infty}^{\infty} J_k{}^2(n\beta) = 1, \forall n\beta \tag{6.19}$$

The spectrum amplitude $\xi(f, \beta)$ of driving pulse $\xi(t)$ at frequency point f is

$$\xi(f,\beta) = \sum_{n=-\infty}^{\infty} C_n \left\{ J_0(n\beta)\,\delta(f - nf_s) + \sum_{k=1}^{\infty} J_k(n\beta) \left[\delta\left(f - nf_s - kf_m\right) + (-1)^k \delta\left(f - nf_s + kf_m\right)\right] \right\} \tag{6.20}$$

So it can be obtained that the power spectrum $S_\xi(\omega)$ in the periodic spread frequency PWM pulse process is

$$\begin{aligned} S_\xi(\omega) &= \lim_{T\to\infty} \frac{1}{T} |\xi(f,\beta)|^2 \\ &= 2\left| \sum_{n=1}^{\infty} C_n \left\{ J_0(n\beta)\,\delta(f - nf_s) + \sum_{k=1}^{\infty} J_k(n\beta) \begin{bmatrix} \delta\left(f - nf_s - kf_m\right) \\ +(-1)^k \delta(f - nf_s + kf_m) \end{bmatrix} \right\} \right| \end{aligned} \tag{6.21}$$

If $\beta = 0$ and $J_0(0) = 1$ in Equations 6.18 and 6.19, the normal fixed-frequency PWM pulse spectrum can be expressed by

$$\xi(f, \beta) = \sum_{n=-\infty}^{\infty} C_n \delta(f - nf_s) \tag{6.22}$$

And the power spectrum density is

$$S(f) = 2\sum_{n=1}^{\infty} |C_n|^2 \delta(f - nf_s) \tag{6.23}$$

The power spectrum only contains the harmonic frequency of integer frequency f_s.

About the periodic spread spectrum PWM, the higher or lower sideband centered in integer multiples f_s is brought by $J_k(n\beta)$ based on Equations 6.17 and 6.18. Thus, the bandwidth B_n of the n-order harmonic is

$$B_n = 2(n\beta + 1)f_m \tag{6.24}$$

The power spectrums of the fixed carrier and sinusoidal modulation carrier are shown in Figure 6.4. The dashed line stands for the fixed carrier power spectrum, and the active line stands for the sinusoidal modulation carrier power spectrum. We can find that the amplitude of the frequency spectrum spreads to more frequency points after the sinusoidal modulation.

From Equation 6.23, the PWM driving pulse power spectrum focuses on the integer frequency points of f_s, which can be proved by the fixed PWM driving pulse waves. The fixed PWM driving pulse waves and their spectrum distribution are shown in Figures 6.5a and 6.6a respectively.

The waves of the fixed-frequency PWM drive pulse are shown in Figure 6.5a. In sinusoidal spread spectrum (SPSP), the PWM driving frequency value is periodically changing during the sinusoidal signal cycle of 0.1 ms, and the varying pattern is decided by sinusoidal SPSP signal as shown in Figure 6.5b. The state of the sawtooth wave SPSP is shown in Figure 6.5c. The PWM pulse frequency changes from small to large during every sawtooth SPSP signal cycle of 0.1 ms, and it changes synchronously

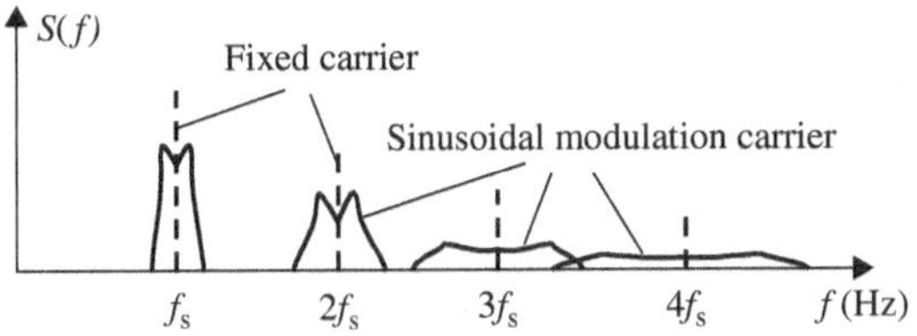

Figure 6.4 Carrier frequency modulation [8].© [2002] IEEE. Reprinted, with permission, from Tse, K. K, Chung, H.S-H, Hui, S. Y., and So, H. C., A comparative study of carrier-frequency modulation techniques for conducted E MI suppression in PWM converters. *IEEE Trans. Industry Electronics*, 2002

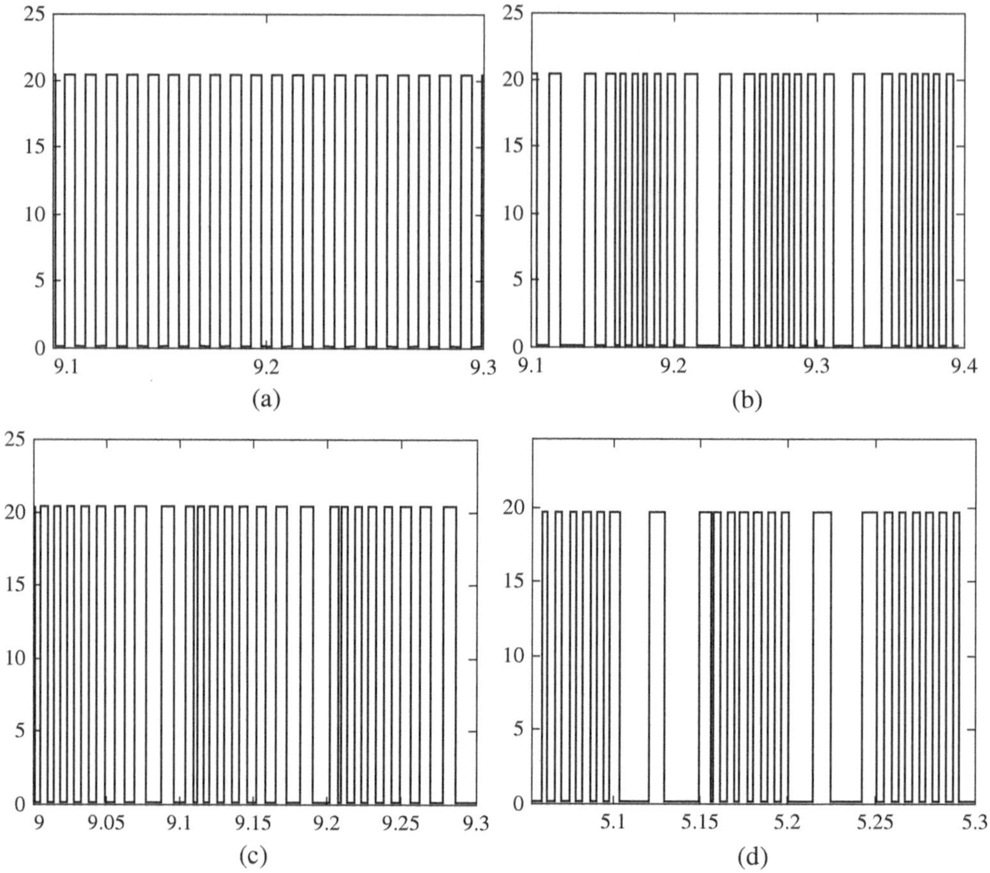

Figure 6.5 PWM drive waveform with (a) fixed frequency PWM, (b) sinusoidal signal spread spectrum, (c) sawtooth signal spread spectrum, and (d) square signal spread spectrum

with the sawtooth wave SPSP signal. As shown in Figure 6.5d, the frequency value variation of the metal oxide semiconductor field effect transistor (MOSFET) drive pulse is decided by the square SPSP signal voltage during the square signal cycle every 0.1 ms by using square wave modulation. Therefore, the effectiveness of PWM frequency control strategy Equation 6.3 has been proved by Figure 6.5.

PWM driving pulse SPSPs consist of infinitely many discrete spectrums, and are decided by β and f_m. Although the periodic spread frequency can reduce the peak value of PWM pulse SPSP, it will still obtain a discrete SPSP and the power focuses on special frequency points centered by $nf_s \pm kf_m$, not spread completely. For the three periodic SPSP methods, the spectrum of sawtooth SPSP is smoother, sinusoidal SPSP takes second place, and square SPSP is the most discrete type as shown in Figure 6.6.

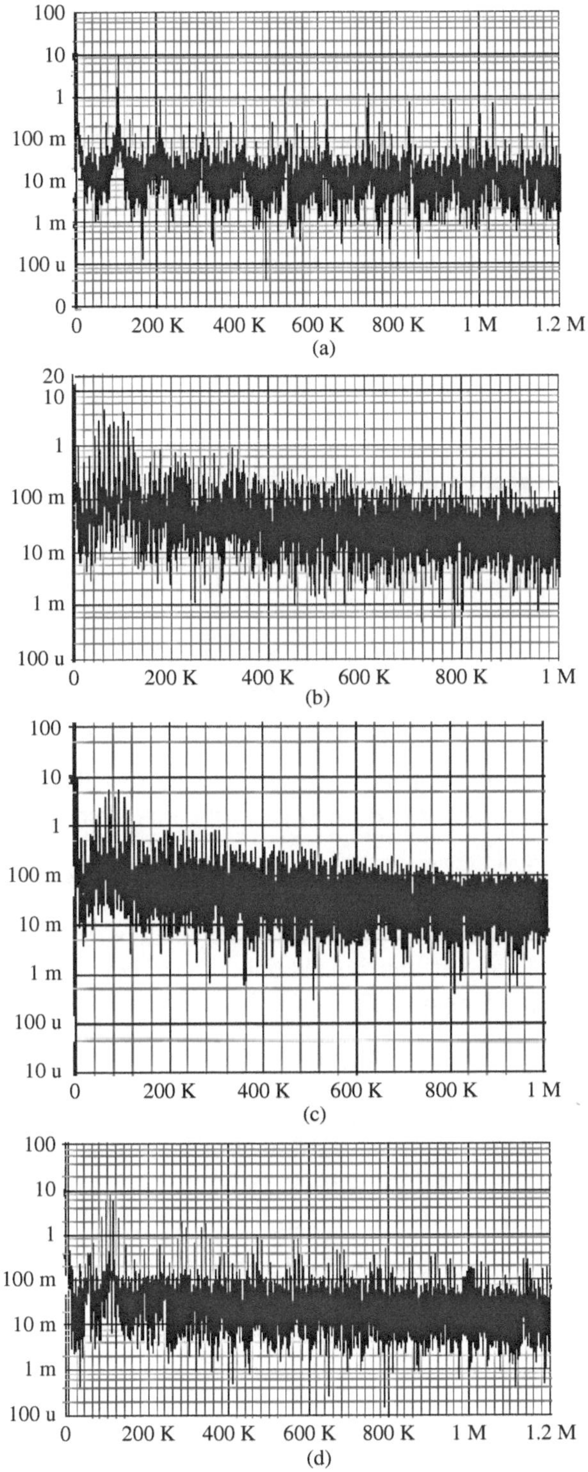

Figure 6.6 Spectrum of the PWM drive waveform with (a) fixed-frequency PWM, (b) sinusoidal signal SPSP, (c) sawtooth wave signal SPSP, and (d) square signal SPSP

6.3.2 Spectral Quantification Analysis of PWM Chaotic SPSP

From Figure 6.2, the PWM pulse interval T_k controlled by chaotic SPSP can be described by the following chaotic mapping φ

$$T_{k+1} = \varphi(T_k) \tag{6.25}$$

The driving pulse starting time τ_k of number K is a chaotic sequence, so

$$\tau_{k+1} = T_1 + \varphi(T_1) + \varphi^{(2)}(T_1) + \cdots + \varphi^{(K)}(T_1) = \phi^{(k)}(T_1) \tag{6.26}$$

Then, the joint probability density function of the chaotic sequence is

$$f_{\tau_1,\cdots\tau_k}(x_1,\cdots,x_k) = \frac{\partial^k F(\tau_1 < x_1,\cdots \tau_k < x_k)}{\partial x_1 \cdots \partial x_k} = f_\tau(x_1)\cdot\prod_{i=1}^{k-1}\delta\left(x_{i+1} - \phi^{(i+1)}(x_1)\right) \tag{6.27}$$

The characteristic function of corresponding τ_k is

$$\begin{aligned}\Theta_{\tau_{1,\ldots,}\tau_K}(\omega_1,\cdots,\omega_k) &= E\left(\exp\left(j\sum_{i=1}^{k}\omega_i\cdot\tau_i\right)\right)\\ &= \int_x\cdots\int\left(\exp\left(j\sum_{i=1}^{k}\omega_i\cdot x_i\right)\right)f_{\tau_{1,\ldots,}\tau_K}(x_1,\cdots,x_k)\end{aligned} \tag{6.28}$$

The probability density function $f_\tau(x)$ of τ_k is an invariant distribution of chaotic mapping φ, and the convergence of invariant distribution is decided by the Frobenius–Perron operator [9]. Here, the Frobenius–Perron operator expressed by $\left(\rho(y) = \sum_{x_i=f^{-1}(y)}\frac{\rho(x_i)}{f'(x_i)}\right)$ is linear and in infinite dimensions. The invariant distribution is the fixed points of Frobenius–Perron operator or the eigenfunction with the eigenvalue 1, and it can be obtained by iterating the Frobenius–Perron operator from the initial density $f_0(x)$

$$f_{k+1}(x) = \rho f_k(x) \tag{6.29}$$

For any fixed time T, it satisfies $\tau_{N(T)} \le T \le \tau_{N(T)+1}$. $N(T)$ is a counted process of the chaotic PWM pulse driving process $\xi(t)$ defined by Equation 6.8. Then, the Fourier spectrum of the chaotic PWM driving process $\xi(t)$ is

$$\xi_T(j\omega) = \sum_{i=1}^{N(t)} Ae^{-j\omega\tau_{i-1}} \tag{6.30}$$

Thus, the power spectrum density $S_\xi(\omega)$ of the chaotic SPSP PWM pulse process is [5]

$$S_\xi(\omega) = \lim_{T\to\infty} \frac{1}{T} E(|\xi_T(j\omega)|^2) \tag{6.31}$$

$$|\xi_T(j\omega)|^2 = \sum_{i=1}^{N(T)} \sum_{k=1}^{N(T)} Ae^{-j\omega\tau_i}(Ae^{-j\omega\tau_k}) = \sum_{k=1}^{N(T)} A^2 + 2\sum_{i=1}^{N(T)-1} \sum_{k=1}^{N(T)-1} A^2 \cos(\omega(\tau_{i+k} - \tau_i)) \tag{6.32}$$

The mathematical expectation can be obtained as follows

$$E(|\xi_T(j\omega)|^2) = NA^2 + 2\sum_{L=1}^{\infty} P_{N(T)}(T) \sum_{i=1}^{N(T)-1} \sum_{k=1}^{N(T)-1} A^2 \cos(\omega(\tau_{i+k} - \tau_i)) \tag{6.33}$$

Here, $N = E(N(T))$ and $P_{N(T)}(T)$ is the probability of event $\{\tau_{N(T)} \le T \le \tau_{N(T)+1}\}$.

Defining $\varepsilon = T - \tau_{N(T)}$, T is the fixed time as mentioned before, and $\tau_{N(T)}$ is the arrival time of the number $N(T)$ chaotic driving pulse. Thus, the probability density function of ε is

$$p_\varepsilon(x) = \sum_{k=1}^{\infty} F_k(T) p_{\tau_k}(T - x) \tag{6.34}$$

where p_{τ_k} is the invariant distribution of starting time τ_k, which is the number k chaotic PWM driving pulse, and $F_k(T)$ is the discrete probability distribution of $N(T)$.

When $T\to\infty$, the probability density function of ε is

$$p_\varepsilon(x) = \frac{1}{E(T_k)} \int_x^{\infty} p_{\tau_k}(y)dy \tag{6.35}$$

According to Wald's law, one can get

$$E(T) = E(\varepsilon + \tau_{N(T)}) = E(\varepsilon) + E(\tau_{N(T)}) \approx \varepsilon' + N \cdot E(T_k) \tag{6.36}$$

For the reason that

$$0 \le \varepsilon' = E(\varepsilon) << NE(T_K)$$

So therc is

$$E(T) \approx NE(T_k) \tag{6.37}$$

By Equations 6.33 and 6.28, the power spectrum density of the chaotic PWM pulse process in Equation 6.26 can be expressed as follows

$$S_\xi(\omega) = \lim_{T\to\infty} \frac{1}{NE(T_K)} A^2 E\left(N + 2\sum_{L=1}^{\infty} P_{N(T)}(T) \sum_{i=1}^{N(T)-1} \sum_{k=1}^{N(T)-1} \cos(\omega(\tau_{i+k} - \tau_i))\right) \tag{6.38}$$

because $\lim_{T\to\infty} \lim_{N(T)\to\infty} P_{N(T)}(T) = 1$.

Using the Birkhooff Ergodic Theorem and Lebesque measurement [9, 10], the further obtained power spectrum density of the PWM pulse is

$$S_\xi(\omega) = \frac{1}{E(T_k)} A^2 \sum_{k=0}^{\infty} \int f_{\tau_1,\cdots\tau_K}(x_1, \cdots, x_k) E\left(\cos \omega \sum_{j=1}^{k} x_j\right)$$
$$= \frac{1}{E(T_k)} \sum_{k=0}^{\infty} A^2 \Theta_{\tau_k}(\omega) \tag{6.39}$$

where $\Theta_{\tau_k}(\omega)$ is the characteristic function of the chaotic PWM driving pulse arrival time τ_k.

From Equations 6.38 and 6.39, the spectrum of the chaotic SPSP controlled PWM pulse which has a continuous chaotic SPSP driving pulse power spectrum is no longer fastened on the special frequency points. However, it contains the spectrum of every

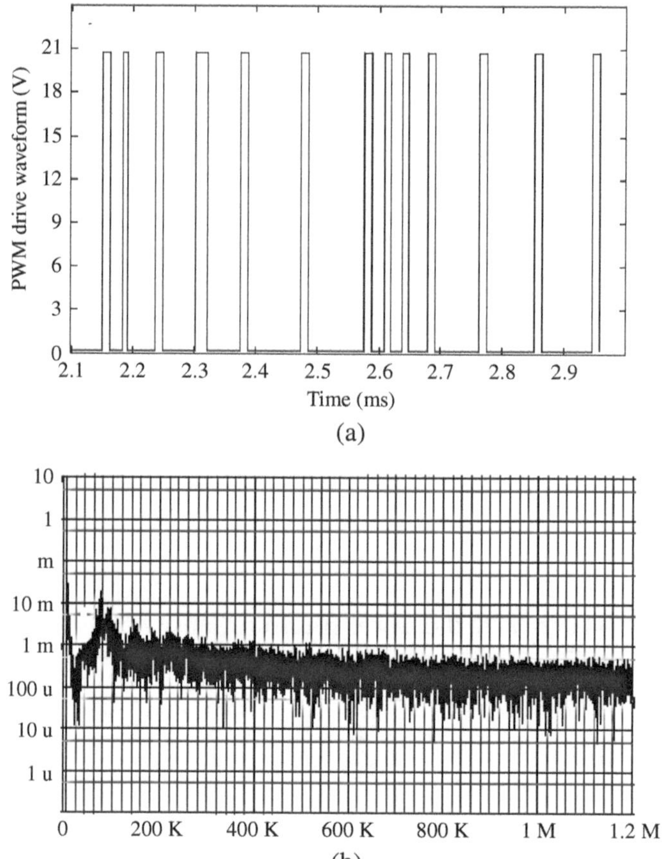

Figure 6.7 (a) PWM driving wave and (b) PWM driving spectrum under chaotic frequency-spreading

frequency point. On the other hand, the power no longer focuses on special frequency points, but is spread throughout the entire frequency. Therefore, the EMI suppressing effect of chaotic SPSP is much better than periodic SPSP, and it has become an effective way to solve converter EMI problems. The chaotic SPSP controlled PWM pulse and spectrums are shown in Figure 6.7; it can be seen that the spectrum is continuous and the peak is smooth.

6.4 Conclusions

Taking an AC–DC converter as an example to analyze the spectrum characteristics of converter SPSP PWM, this chapter studied the PWM driving pulse. It has been proved in theory that the EMI peak of the fixed frequency PWM is higher and cannot be suppressed effectively because of the discrete spectrum and high frequency EMI focused on the frequency multiplication area. By putting a PWM driving wave as a process or event flow, we can obtain conclusions based on Fourier spectrum and k order Bessel function analysis. First, although periodic PWM can change the spectrum distribution, it cannot change the characteristics of the discrete spectrum distribution and limits the suppressing PWM effect. Second, using random point process theory based on the principle of statistics has allowed analysis of the characteristic function of chaotic SPSP PWM and its probability density function. And the power spectrum density of chaotic SPSP PWM has been analyzed through invariant distribution calculations. Finally, it turns out that the spectrum of chaotic SPSP PWM is fully continuous. So it can suppress the converter EMI effectively and has huge potential for suppressing EMI.

References

[1] Snider, D.L. (1982) *Stochastic Point Processes*, People's Education Press, Beijing.

[2] Deng, Y. and Liang, Z. (1992) *Stochastic point processes and applications*, Science Press, Beijing (in Chinese).

[3] Hu, Z., Zhang, B., Hu, S. *et al.* (2005) The study of DC-DC converter output controllability based on switched linear system (in Chinese). *Chin. J. Electron.*, **33**(2), 370–373.

[4] Yang, R., Zhang, B. and Qiu, D.Y. (2008) Chaotic point proce s s de scription of converter discrete subsystem and EMI suppression. *Acta Phys. Sinica*, **57**(3), 1389–1397.

[5] Baranovski, A.L. and Schwarz, W. (2003) Chaotic and random point processes: analysis, design, and applications to switching systems. *IEEE Trans. Circuits Syst. I*, **50**(8), 1081–1088.

[6] Yang, R., Zhang, B. and Qiu, D.Y. (2006) Quantification de scription of the chaotic PWM spectrum of the switching converter. *Acta Phys. Sinica*, **55**(11), 5667–5673.

[7] Carlson, A.B. (1986) *Communication Systems: An Introduction to Signal and Noise in Electrical Communication*, McGraw-Hill Press, New York.

[8] Tse, K.K., Chung, H.S.-H., Huo, S.Y. and So, H.C. (2002) A comparative study of carrier- frequency modulation techniques for conducted E MI suppression in PWM converters. *IEEE Trans. Ind. Electron.*, **49**(3), 618–627.

[9] Ott, E. (2002) *Chaos in Dynamical Systems*, Cambridge University Press, London.

[10] Gao, P. (2005) *Nonlinear Dynamics*, NUDT Press, China (in Chinese).

7

Basis of Spectral Analysis

7.1 Introduction

A wide spectrum of harmonic components is generated by high frequency power converters, which deteriorates the quality of the delivered energy, increases the energy losses, and decreases the reliability of a power system [1].

In many practical problems (in particular studies of nonlinear circuits), only a discrete time series $x(n)$, not a continuous signal $x(t)$, is given. Therefore, how to extract the state information from these time series becomes a serious problem. Fortunately, the characteristic behavior is easy to distinguish by spectrum analysis which is mainly concerned with estimating the spectrum over the whole range of frequency. Essentially, spectrum analysis is a modification of Fourier analysis so as to make it be suitable for stochastic rather than deterministic functions of time.

In Chapter 1, we saw that periodic motion corresponds to a spike in the power spectrum, and the characteristics of chaos motion are of noise background and a wide spike in the power spectrum. And we also saw that switching converters generate not only the characteristic harmonics typical for the ideal converter operation, but also a considerable amount of noncharacteristic harmonics and inter-harmonics which may strongly deteriorate the quality of the power supply voltage [2]. Inter-harmonics is defined as the noninteger harmonics of the main fundamental under consideration. Estimation of spectrum, including estimating the strength of the electromagnetic interference (EMI) of converters via measuring the system harmonics distribution and studying the nonlinear behavior of converters via spectral analysis, is useful in power electronics. Intuitively speaking, the spectrum characterizes the frequency contents of the signal. The goal of power spectral analysis is to describe the distribution (over frequency) of the power contained in a signal based on a continuous signal or a finite set of data.

In order to deal with spectrum in a proper scientific manner, in this chapter we return to the main theme of spectral analysis. We will discuss this topic with the Fourier transform first, followed by the concept of power spectral density (PSD). Then, the classic and modern PSD estimation methods will be introduced.

Chaos Analysis and Chaotic EMI Suppression of DC-DC Converters, First Edition. Bo Zhang and Xuemei Wang.

7.2 Some Concepts

Understanding the basic concepts and basic characteristics of signals and systems will help us to follow the more concrete spectral analysis. For power electronic engineers, acquiring spectrum analysis tools is more practical than theories. So, we will introduce these basic definitions physically instead of mathematically in the following.

For this study of signals and systems, we will divide signals into two groups: those that have a fixed behavior (deterministic signals) and those that change randomly (stochastic signals). In deterministic signals, each value is fixed and can be determined by a mathematical expression, rule, or table. Because of this, future values of any deterministic signal can be calculated from past values. For this reason, these signals are relatively easy to analyze as they do not change, and we can make accurate assumptions about their past and future behavior. Unlike deterministic signals, stochastic signals, or random signals, are not so nice. Random signals cannot be characterized by a simple, well-defined mathematical equation, and their future values cannot be predicted. Rather, we must use probability and statistics to analyze their behavior. Also, because of their randomness, average values from a collection of signals are usually studied rather than one individual signal. Deterministic functions form the whole domain of study in classical mathematical analysis, but almost all the processes are of this type in practical applications of spectral analysis [3].

As mentioned above, in order to study the nature of random signals, we want to look at a collection of these signals rather than just one instance of that signal. This collection of signals is called a random process.

Definition 7.1 Statistical process A statistical process (or random process) is a random function, or an indexed family of random variables more precisely.

From the definition of a statistical process, we know that all random processes are composed of random variables, each at its own unique point in time. So statistical processes have all the properties of random variables, such as mean, correlation, variances, and so on. To deal with groups of signals or sequences, it will be important for us to be able to show whether or not these statistical properties hold true for the entire statistical process. To do this, the concept of stationary processes has been developed.

Definition 7.2 Stationary processes A random process is its statistical properties and these do not vary with time.

Processes whose statistical properties do change are referred to as **nonstationary**.

Stationarity is a basic assumption in classical time series analysis. A stationary signal is independent of time, for example, a DC signal and $\cos(20\pi t) + \cos(50\pi t)$, whose frequency components don't change over time. Conversely, frequency components of a nonstationary signal change over time. For example, the frequencies of human speech vary over time and depend on the words or syllables pronounced. Generally, wide-sense stationary is used to characterize the signal; for a detailed description of wide-sense stationary please refer to reference [4].

Definition 7.3 Resolution Resolution refers to the ability to discriminate spectral features, and is a key concept in the analysis of spectral estimator performance.

7.3 Fourier Analysis and Fourier Transform

There are a 100 years of research and application of Fourier analysis. Fourier analysis is basically concerned with approximating a function by a sum of sine and cosine terms. It is called Fourier series which is an expansion of a periodic function $f(t)$ in terms of an infinite sum of sines and cosines. The computation and study of Fourier series is known as harmonic analysis and is extremely useful as a way to break up an arbitrary periodic function into a set of simple terms. These terms can be plugged in, solved individually, and recombined to obtain the solution to the original problem or an approximation to it to whatever accuracy is desired or practical.

In general, any periodic function $f(t)$ that is integrable on $[-\pi, \pi]$, may be expressed as a Fourier series

$$f(t) = \sum_{k=-\infty}^{+\infty} a_k e^{ik\left(\frac{2\pi}{T}\right)t} \tag{7.1}$$

where T is the function's period (or integration interval), a_k may be calculated by

$$a_k = \frac{1}{T}\int_T f(t) e^{-ik\left(\frac{2\pi}{T}\right)t} dt \tag{7.2}$$

For a given signal $f(t)$, there are lots of methods to describe it, such as a function expression, a data series, and even a simple timing diagram. But all of these are expressions of the time domain. As we all know, the Fourier transform is used to analyze the spectra of signal $f(t)$, as $F(\omega)$. By Fourier transform, two basic physical quantities, time and frequency, are connected. That is to say, the frequency domain expression is derived by the time domain expression, and vice versa. Furthermore, the power spectrum may be obtained by $F(\omega)$, which is the frequency domain expression of a signal.

Fourier transform has been applied to stationary signals or stationary processes. Moreover, the statistical property of Fourier transform does not change over time, and generally arises from a stable physical system which has achieved a steady-state mode. In mathematics, a stationary process is a stochastic process whose joint probability distribution does not change when shifted in time or space.

Fourier analysis is extremely useful for data analysis, as it breaks down a signal into constituent sinusoids of different frequencies. Mathematically, the process of Fourier analysis is represented by the Fourier transform

$$F(\omega) = \int_{-\infty}^{\infty} f(t) e^{-j\omega t} dt \tag{7.3}$$

Fourier transform is the sum over all time of the signal $f(t)$ multiplied by a complex exponential which can be broken down into real and imaginary sinusoidal components.

The results of the transform are the Fourier coefficients which multiplied by a sinusoid of a specific frequency yield the constituent sinusoidal components of the original signal. Graphically, this process looks like Figure 7.1.

In Figure 7.1, the original signal is $f(t) = 2\sin[2\pi(50t)] + \sin[2\pi(120t)]$, and it can be decomposed into two sine functions with different frequencies by Fourier transform.

The sequence $x(n)$ is usually the result of sampling a continuous signal $x(t)$ at some uniform sampling rate Fs. The sequence $x(n)$ is called a discrete sequence $x(n)$. If $x(n)$ is absolutely summable, namely, $\sum_{n=-\infty}^{\infty} |x(n)| < \infty$, then the discrete time Fourier transform (DTFT) of a series $x(n)$ is given by

$$X(\omega) = \sum_{n=-\infty}^{\infty} x(n)e^{-j\omega n} \tag{7.4}$$

where $\omega \in (-\pi, +\pi)$ denotes the continuous radian frequency variable. $X(\omega)$ is a function of angular frequency ω, so the DTFT frequencies form a continuum. Compared with the Fourier transform $F(\omega)$ of an analog signal, $X(\omega)$ similarly presents the distribution law in the frequency domain except the signal whose period is 2π.

The DTFT provides the frequency domain representation for absolutely summable sequences. This transform commonly has two features. First, the transform is defined for infinite length sequences. Second, it is a function of continuous variable ω. In other words, the DTFT is not a numerically computable transform. Therefore, a numerically computable transform should be developed. Practically, a new transform called the discrete Fourier transform (DFT) for a finite-duration sequence is defined as

$$X(\omega_k) = \sum_{n=0}^{N-1} x(t_n)e^{-j\omega_k t_n} \quad k = 0, 1, \cdots, N-1 \tag{7.5}$$

where T is the sampling period, N is the number of samples, $t_n = nT$ is the nth sampling instant, and $\omega_k = 2\pi k/N$ denotes the kth frequency sampled. So, $X(\omega_k)$ denotes the spectrum of x at frequency ω_k, and the DFT is a function of discrete frequency ω_k.

It is common to rewrite the DFT in the more pure mathematic form obtained by setting $T = 1$ in the previous definition

$$X(k) = \frac{1}{N}\sum_{n=0}^{N-1} x(n)e^{-j\frac{2\pi}{N}kn} \quad k = 0, 1, \cdots, N-1 \tag{7.6}$$

Figure 7.1 Schematic diagram of Fourier transform

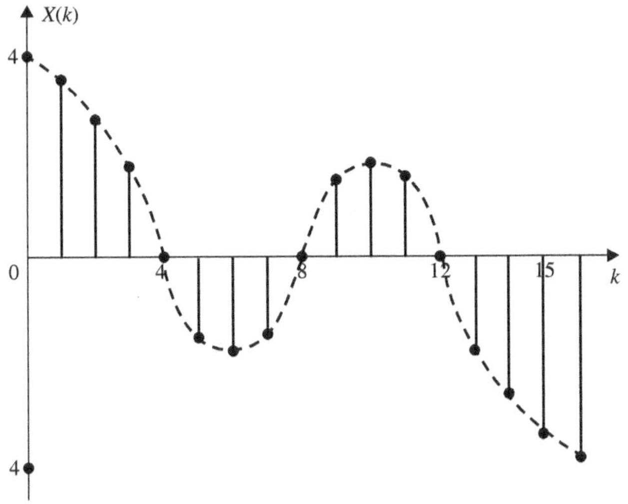

Figure 7.2 Amplitude spectrums of DFT and FT of a sampled sequence at $k = 16$

$X(k)$ denotes the kth spectral sample, and it is named the N-point DFT of $x(n)$. This form is the simplest mathematically, while the previous form is easier to interpret physically.

The physical significance of DFT can be illustrated as: DFT is the N-point sample at equal intervals of DTFT. Figure 7.2 describes the relationship of DFT and DTFT clearly. In this figure, the amplitude spectrum of the DTFT $X(\omega)$ and DFT $X(k)$ are shown in dashed and vertical lines at equal intervals respectively.

DFT is one of the most important transforms in signal analysis and processing. The computation of DFT is proportional to N, so the amount of calculation will be quite large if N is large. This resulted in the DFT algorithm being unrealistic in spectrum analysis and real-time signal processing until the fast Fourier transform (FFT) algorithm emerges. The FFT, devised in the mid-1960s, is an efficient algorithm for computing the DFT of a series with reduced execution time. It is not a separated transform and is particularly useful in areas such as signal and image processing, filtering, convolution, frequency analysis, and power spectrum estimation.

The Fourier method gives a frequency point of view of a signal. The signal is considered as a sum of sinusoids. The problem is that a sinusoid never begins and never stops. The sinusoid function is always the same from minus infinite to infinite. So it is very difficult to locate when a frequency is present. However, it is sometimes possible because all the information is present in the Fourier transform [5]. Four cases of Fourier transform are shown in Table 7.1.

7.4 Spectral Density

In the last section, we noticed that periodic and deterministic nonperiodic functions could be expressed as sums of sine and cosine terms with different amplitudes and

Table 7.1 Four cases of sampled/continuous finite/infinite time and frequency of Fourier transform

Time duration		
Infinite	Finite	
Fourier transform (FT) $F(\omega) = \int_{-\infty}^{\infty} f(t)e^{-j\omega t}dt$ $\omega \in (-\infty, +\infty)$	Fourier series (FS) $F(k) = \int_{0}^{P} f(t)e^{-j\omega_k t}dt$ $k = -\infty, \cdots, +\infty$	Continuous time t
Discrete time FT (DTFT) $X(\omega) = \sum_{n=-\infty}^{\infty} x(n)e^{-j\omega n}$ $\omega \in (-\pi, \pi)$	Discrete FT (DFT) $X(k) = \sum_{n=0}^{N-1} x(n)e^{-j2\pi kn/N}$ $k = 0, 1, \cdots, N-1$	Discrete time n
Continuous frequency ω	Discrete frequency k	

frequencies. The importance of this so-called "spectral representation" of a function lies in the fact that if the function represents some physical process, such as a current or voltage, the total energy dissipated by the process in any time interval is equal to the sum of the amounts of energy dissipated by each of the sine or cosine terms [6]. In addition, spectral density represents the contribution of every frequency component of the power of the overall signal, and it plays an important role in studies of the distributions of signal energy/power over frequency. Especially, the random signals, which may not be expressed by deterministic functions, could not be described by a spectrum. Therefore, the energy/PSD as speech signal recognition, radar clutter analysis, and seismic signal processing will be introduced in this section. The goal of spectral density estimation is to estimate the spectral density of a random signal from a sequence of time samples of the signal. Intuitively speaking, the spectral density characterizes the frequency content of the signal. The purpose of estimating spectral density is to detect any periodicities in the data, by observing peaks at the frequencies corresponding to these periodicities.

7.4.1 Energy Signals and Power Signals

The mathematical methods employed in the analysis of signals depend on the characteristics of signals. The definitions of energy and power signals will be introduced in this section first.

The normalized energy of signal $f(t)$ is defined as the energy consumption of a signal's voltage or current applied to a 1 Ω resistance

$$E = \int_{-\infty}^{\infty} |f(t)|^2 dt \tag{7.7}$$

For the discrete time signals $x(n)$, one can obtain

$$E = \sum_{n=-\infty}^{\infty} |x(n)|^2 \tag{7.8}$$

Usually, the signal with finite energy is called the energy signal. In practice, the common nonperiodic signal belongs to a finite energy signal. However, signals like the periodic signal are infinite, so its energy may not be studied again.

The signal's average power is defined as the power consumption of signal voltage or current flows through a 1 Ω resistance. The average power of $f(t)$ in time interval $[T_1, T_2]$ is expressed as

$$P = \frac{1}{T_2 - T_1} \int_{T_1}^{T_2} |f(t)|^2 dt \tag{7.9}$$

If the signal's power is finite, this signal is called a power signal.

Moreover, the average power of a signal $f(t)$ in the whole timeline $[-\infty, \infty]$ is

$$P = \lim_{T \to +\infty} \frac{1}{T} \int_{-\frac{T}{2}}^{\frac{T}{2}} |f^2(t)| dt \tag{7.10}$$

So the power of a signal may be finite even if the energy is infinite.

For the discrete time signals $x(n)$, there is

$$P = \lim_{N \to \infty} \frac{1}{2N+1} \sum_{n=-N}^{N} |x(n)|^2 \tag{7.11}$$

Specially, if $x(n)$ is a periodic signal with fundamental period N and takes on finite values, its power is given by

$$P = \frac{1}{N} \sum_{n=0}^{N-1} |x(n)|^2 \tag{7.12}$$

Consequently, periodic signals are power signals.

The Fourier transform of a power signal does not exist because it is not absolutely integrable. For example, the Fourier transform of a sine function is nonexistent and only the introduction of an impulse function may obtain its Fourier transform. Thus, the power signal is always cut off as in Figure 7.3 because the Fourier transform cannot use it directly.

7.4.2 Energy Spectral Density

Energy spectral density describes how the energy of a signal or a time series is distributed with frequency. The energy spectral density is most suitable for transients; for example, pulse-like signals have a finite total energy. In this case, based on the

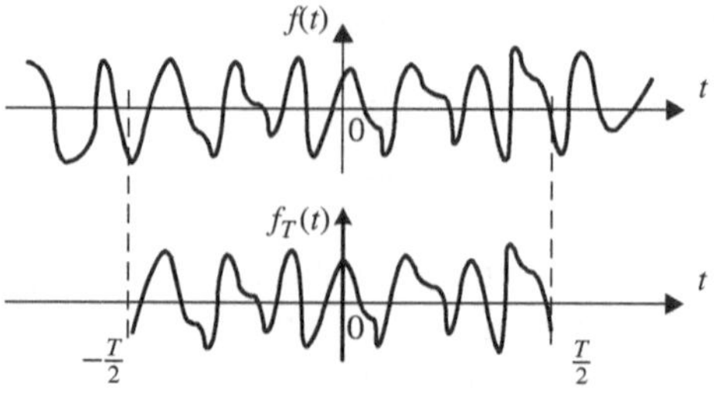

Figure 7.3 Power signal and after cut-off

Parseval theorems [7], we have an alternate expression for the energy of the signal in terms of its Fourier transform

$$E = \int_{-\infty}^{\infty} f^2(t)dt = \frac{1}{2\pi}\int_{-\infty}^{\infty} |F(\omega)|^2 d\omega \tag{7.13}$$

Defining

$$E(\omega) = |F(\omega)|^2 \tag{7.14}$$

where $E(\omega)$ represents the distribution of a signal's energy in the frequency domain, is called energy spectrum density (ESD). $E(\omega)$ reflects the energy per unit frequency contained in the signal at frequency ω, and its unit is J/H. For the discrete signal $x(n)$, the $F(\omega)$ is replaced by $X(\omega)$.

7.4.3 Power Spectral Density

The energy spectral density is most suitable for transient, that is, pulse-like signals, for which power is limited and the Fourier transform is presented.

For deterministic signals, the Fourier transform may be used to analyze their spectral characteristics. For wide-sense stationary stochastic signals, however, their Fourier transforms do not exist strictly because they are not only periodic but also satisfy the condition of squared integrable. So the PSD is always used to analyze their spectrum.

For continuous signals such as sine signals and square wave signals, it makes more sense to define a PSD, which describes how the power of a signal or time series is distributed over the different frequencies. Here, power can be the actual physical power, or can be defined as the squared value of the signal for convenience with abstract signals such as Equation 7.9.

In analyzing the frequency distributions of the signal $f(t)$, one likes to compute the ordinary Fourier transform $F(\omega)$. However, we are interested in the fact that many signals have no Fourier transforms. Because of this, it is advantageous to work with a truncated Fourier transform, where the signal is integrated only over a finite interval $[0, T]$.

It can be proven

$$P = \lim_{T\to+\infty} \frac{1}{T}\int_{-\infty}^{\infty} f^2(t)dt = \frac{1}{2\pi}\int_{-\infty}^{\infty} \frac{1}{T}|F(\omega)|^2 d\omega$$

where

$$F(\omega) = \int_{-\frac{T}{2}}^{\frac{T}{2}} f(t)e^{-j\omega t}dt \tag{7.15}$$

Then the PSD can be defined as

$$P(\omega) = \lim_{T\to+\infty} \frac{1}{T}|F(\omega)|^2 \tag{7.16}$$

Obviously, PSD describes how the power of a signal or time series is distributed in the frequency domain. And the power is defined as the actual power if the signal is a voltage applied to a 1 Ω resistance. PSD is commonly expressed in watts per hertz (W/Hz).

For the discrete sequence $x(n)$, PSD can be expressed as

$$P(\omega) = \frac{1}{N}|X(\omega)|^2 \tag{7.17}$$

where $X(\omega)$ is the DTFT of $x(n)$.

Apparently, both ESD and PSD retain the magnitude of the spectrum information, but lose phase information. Therefore, the energy/power spectrums of different signals perhaps are the same.

7.5 Autocorrelation Function and Power Spectral Density

If the average value of the signal is non-zero and it is not square integrable, the Fourier transforms do not exist and the PSD cannot be estimated in this case. Fortunately, the Wiener–Khinchin theorem provides a simple alternative. The PSD is the Fourier transform of the autocorrelation function of the signal if the signal can be treated as a wide-sense stationary random process.

The power spectrum of a stationary random process x_n is mathematically related to the correlation sequence by the discrete-time Fourier transform. In terms of normalized frequency, this is given by

$$S_{xx}(\omega) = \sum_{m=-\infty}^{\infty} R_{xx}(m)e^{-j\omega m} \tag{7.18}$$

where $R_{xx}(m)$ is the autocorrelation function of a sequence $x(n)$. The autocorrelation of a signal x is simply the cross-correlation of x with itself

$$R_{xx}(m) = \frac{1}{N}\sum_{n=0}^{N-1} x(n)x(n+m) \tag{7.19}$$

Equation 7.18 is called the Wiener–Khintchine Theorem, which states that the autocorrelation function of a wide-sense stationary random process has a spectral decomposition given by the power spectrum of that process. According to this theorem, the autocorrelation function of a discrete-time zero-mean stationary random signal and its PSD are a pair of DFTs.

Equation 7.18 also can be written as a function of physical frequency f (e.g., in hertz) by using the relation $\omega = 2\pi f/f_s$

$$S_{xx}(f) = \sum_{m=-\infty}^{\infty} R_{xx}(m)e^{-2\pi jfm/f_s} \tag{7.20}$$

where f_s is the sampling frequency. By sampling the discrete-time sequence, the spectral density is periodic in the frequency domain.

The correlation sequence can be derived from the power spectrum by using the Inverse Discrete-time Fourier transform (IDFT):

$$R_{xx}(m) = \int_{-\pi}^{\pi} \frac{S_{xx}(\omega)e^{j\omega m}}{2\pi} d\omega = \int_{-f_s/2}^{f_s/2} \frac{S_{xx}(f)e^{2\pi jfm/f_s}}{f_s} df \tag{7.21}$$

The average power of the sequence $x(n)$ over the entire Nyquist interval is represented by

$$R_{xx}(0) = \int_{-\pi}^{\pi} \frac{S_{xx}(\omega)}{2\pi} d\omega = \int_{-f_s/2}^{f_s/2} \frac{S_{xx}(f)}{f_s} df \tag{7.22}$$

The quantities are

$$P_{xx}(\omega) = \frac{S_{xx}(\omega)}{2\pi} \tag{7.23}$$

and

$$P_{xx}(f) = \frac{S_{xx}(f)}{f_s} \tag{7.24}$$

So, the Fourier Transform of the autocorrelation sequence is defined as the Power Spectral Density of the stationary random signal $x(n)$. You can see from the above expression that $P_{xx}(\omega)$ represents the power content of a signal in an infinitesimal frequency band, which is why it is called the power spectral density.

The average power of a signal over a particular frequency band $[\omega_1, \omega_2], 0 \leq \omega_1 < \omega_2 \leq \pi$ can be found by integrating the PSD over that band

$$\overline{P}_{[\omega_1,\omega_2]} = \int_{\omega_1}^{\omega_2} P_{xx}(\omega)d\omega + \int_{-\omega_2}^{-\omega_1} P_{xx}(\omega)d\omega \tag{7.25}$$

From the above expression, you can see that $P_{xx}(\omega)$ represents the power density of a signal in an infinitesimal frequency band. The unit of the PSD is power (e.g., watts) per unit of frequency. In the case of $P_{xx}(\omega)$, it is watts/radian/sample or simply watts/radian. In the case of $P_{xx}(f)$, the units are watts/hertz. Integration of the PSD with respect to frequency yields units of watts as is expected for the average power.

For real signals, the PSD is symmetric about DC, and thus $P_{xx}(\omega)$ for $0 \leq \omega \leq \pi$ is sufficient to completely characterize the PSD. However, to obtain the average power over the entire Nyquist interval, it is necessary to introduce the concept of the one-sided PSD. The one-sided PSD is given by

$$P_{\text{onesided}}(\omega) = \begin{cases} 0 & -\pi \leq \omega < 0 \\ 2P_{xx}(\omega) & 0 \leq \omega < \pi \end{cases} \tag{7.26}$$

The average power of a signal over the frequency band $[\omega_1, \omega_2], 0 \leq \omega_1 < \omega_2 \leq \pi$ can be computed using the one-sided PSD as

$$\overline{P}_{[\omega_1,\omega_2]} = \int_{\omega_1}^{\omega_2} P_{\text{onesided}}(\omega)d\omega \tag{7.27}$$

7.6 Classic Power Spectrum Estimation

Stationary random signals do not have finite energy and hence do not process a Fourier transform. Such signals have finite average power and thus are characterized by the PSD. We know that the PSD represents the energy distribution of the random signal over frequency and reveals the implied periodic behavior of signals. However, the stationary random signal in practical application is usually of finite length. Thus, the power spectrum estimation is a problem for estimating the true PSD of the original signal via the finite length data.

Power spectrum estimation can be divided into classic power spectrum estimation (nonparametric model spectrum estimation) and modern power spectrum estimation (parameter model spectrum estimation). The classic spectrum estimation methods derived from the PSD definitions with no assumptions on the data-generating process or model (e.g., autoregressive model) are standards for spectrum analysis because the computation is especially efficient if FFT is employed. The unknown data out of workspace is assumed zero (equivalent to windowing) in the classic spectrum estimation methods, including the periodogram, Bartlett, Welch, and BT methods.

7.6.1 Periodogram

The periodogram was one of the earliest statistical tools for studying periodic tendency and the PSD in time series. The periodogram method, also called direct method, originally introduced by Schuster in 1898, directly estimates the PSD, which means computing the squared-magnitude of the DFT of the signal itself. The periodogram method is the earliest classic spectrum estimation method which is still used today except DFT is replaced by FFT.

The PSD and DFT have a relationship as Equation 7.10. This spectral estimation method is periodicity because $|X(k)|$ is N-periodic. Thus, this famous method is called the periodogram method.

The advantage of the periodogram method is the fast DFT algorithm and it can be applied to estimate the PSD. It should be noticed that the periodogram method is applicable to the case of the long signal sequence. Let an original superposed signal be $f(t) = \sin[2\pi(40t)] + 3\sin[2\pi(100t)] + 0.1$ random noise.

Figure 7.4 is the PSD estimation by periodogram, where a 512-point FFT algorithm has been used with sampling frequently $f_s = 400$ Hz. We can see that the two frequency peaks pump clearly, but the spectrum undulates greatly.

For spectrum estimation using a periodogram, if the date length is too long, the spectrum may undulate aggregately or its variance may become bad. And if the date length is too short, the resolution may be worse. Thus, the periodogram method should be modified to reduce variance, and one way to enforce zero variance is through averaging.

7.6.2 Bartlett

Based on the theory of probability and statistics, if the original data with the length N is independently divided into L subsegments with the length $M = N/L$, the variance will be $1/L$ of the original probability estimation.

In order to reduce the fluctuations of PSD estimation curves, M.S. Bartlett put forward an averaging periodogram method, also known as the Bartlett method. This method provides a way to reduce the variance of the periodogram at the cost of resolution compared with standard periodograms [8, 9].

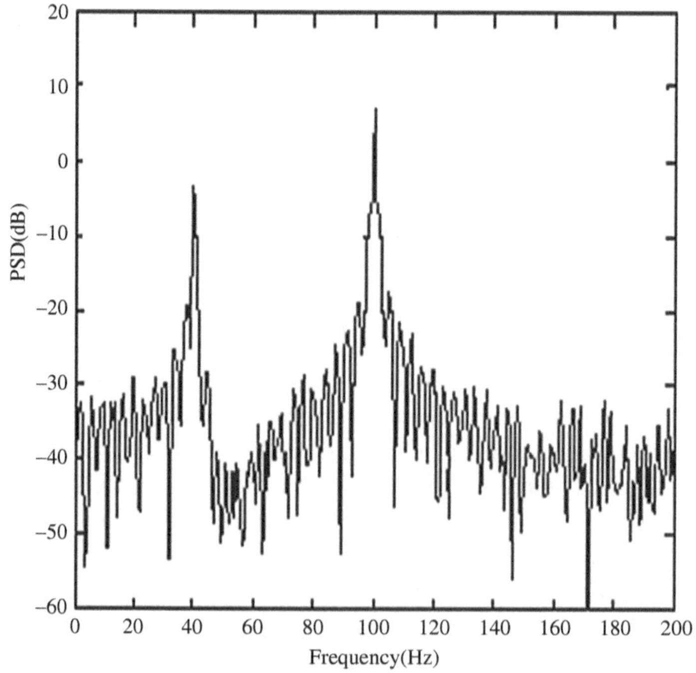

Figure 7.4 PSD estimation via the periodogram method

In the Bartlett method, the sequence x_n is subdivided into L nonoverlapping subsegments where each segment has length M. For each subsegment, the periodogram is computed and the Bartlett power spectrum estimate is obtained by averaging the periodograms for the L segments. Therefore, the variance of this method is reduced to $M = N/L$, and the frequency resolution is reduced by a factor L because of the shortening of the subsegments.

Figure 7.5 is the PSD estimation by Bartlett, where a 512-point FFT algorithm has been used with $f_s = 400\,\text{Hz}$, $L = 8$, and $M = 50$. We can see that the two frequency peaks are 40 and 100 Hz, and the variance of the estimation is significantly reduced.

7.6.3 Welch

In 1967, P.D. Welch proposed another modified periodogram method by windowing and averaging, which means windowing each subsegment before estimating its PSD, and then averaging them [10]. This combination of windowing and averaging may obtain a better PSD estimation of sufficient spectral resolution and lower random fluctuations.

The Welch method made two modifications to the Bartlett method.

1. It allows overlap of subsegments (gives more subsegments, and thus it is "better" averaging).
2. Each data segment is windowed before computing its periodogram.

The two modifications have three merits. The first one is no negative spectrum again, no matter what kind of window function is used. The second one is that the overlapping can further reduce the periodogram variance. The third one is that the windowing can reduce the spectral leakage associated with finite observable duration. Compared with the periodogram, the Welch method may improve the smoothness of the spectrum curves and the resolution of spectrum estimation greatly. In contrast, although the

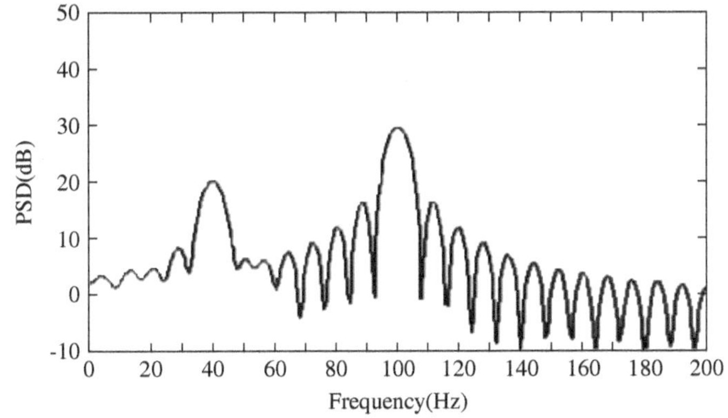

Figure 7.5 PSD estimation via the Bartlett method

PSD produced by Welch's method has wider peaks (called wide main lobes, caused by Gibbs phenomenon [11]), you can still distinguish the two sinusoids, which stand out from the "noise floor."

Figure 7.6 is the PSD estimation by the Welch method without and with windows respectively, where a 512-point FFT algorithm has been used with $f_s = 400$ Hz, $L = 8$, $M = 50$, and overlap = 50%.

7.6.4 Blackman and Tukey Method

By using the Wiener–Khinchin theorem, the power spectrum of a WSS random process is the Fourier transform of the autocorrelation sequence. In 1958, Blackman and Tukey developed the spectrum estimation based on this theorem [12], in which the sampled autocorrelation sequence is windowed first and then the Fourier transform is used to yield the estimation of the power spectrum. This spectrum method is called the BT method, or Correlogram method.

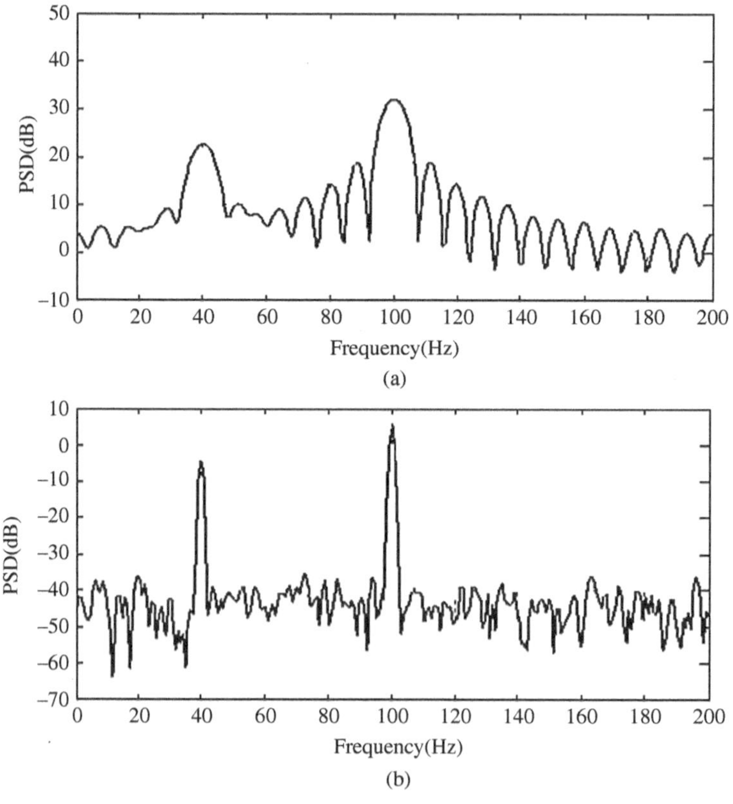

Figure 7.6 PSD estimation via the Welch method (a) unwindowed and (b) windowed by the Hamming window

In the Blackman–Tukey approach, $P_X(f)$ is estimated by

$$P_X(f) = \left| \sum_{k=0}^{M-1} w_k R_k e^{-j\omega k} \right| \tag{7.28}$$

where R_k is the autocorrelation estimate at lag k, M is the maximum lag considered and window length, and w_k is the windowing function. Several window shapes are available: Blackman (rectangle), Bartlett (triangular), Hamming (cosinusoidal), Hanning (slightly differently cosinusoidal), and so on. Figure 7.7 is the PSD estimation using the BT method with Blackman windowing. Obviously, the BT Method has better variance and greater frequency resolution than the Bartlett and Welch methods.

7.6.5 Summary of Classic PSD Estimators

The classical PSD estimation approaches have the following advantages: (i) they are computationally efficient if only a few lags are needed (BT) or if the FFT is used (periodogram), (ii) the PSD estimate is directly proportional to the power for sinusoid processes, and (iii) they area good model for some applications (the model is a sum of harmonically-related sinusoids). The disadvantages of these techniques are (i) suppression of weak signal main lobe responses by strong signal sidelobes, (ii) frequency resolution is limited by the available data record duration, independent of the characteristics of the data or its signal-to-noise ratio (SNR), (iii) introduction of distortion in the spectrum due to side-lobe leakage, (iv) the need for some sort of pseudo ensemble averaging to obtain statistically consistent periodogram spectra, and (v) the appearance of negative PSD values with the BT approach when some autocovariance sequence estimates are used [11].

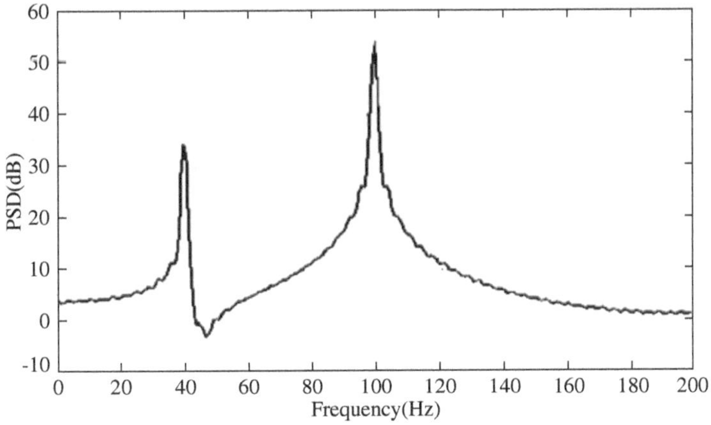

Figure 7.7 BT method with the Blackman window

7.7 Modern Spectral Density Estimation

The nonparametric methods described in the previous sections of this chapter, which utilize periodogram and its modified methods, are subject to the aforementioned limitations of low spectral resolution in the case of short records and the requirement for windowing to reduce the spectral leakage. Considering some invalid assumptions (zero data or repetitive data outside the duration of observation) in these methods, the estimated spectrum can be a smeared version of the true spectrum [13]. Of course, this assumption is not realistic and causes the low resolution of classical PSD estimation. These difficulties may be overcome by parametric methods, termed modern spectral density estimation.

Use of a priori information (or assumptions) may permit the selection of an exact model for the process generating the data samples, or at least a model that is a good approximation to the actual underlying process. It is then usually possible to obtain a better spectral estimate based on the model by determining the parameters of the model from the observations [11]. The main ideas of parametric methods are those in which the PSD is estimated from a signal that is assumed to be the output of a linear system driven by white noise. Next, the parameters (coefficients) of the linear system by the original signal or its autocorrelation function are estimated, then the PSD may be estimated by the parameters of the linear system. These methods estimate the PSD by first estimating the parameters of the linear system that hypothetically "generates" the signal. They tend to yield higher resolutions than classical nonparametric methods when the data length of the available signal is relatively short.

The AR method is a parametric method and is commonly used. Estimating the parameters from AR signal models is a well established topic, and then the estimation can be found by solving the linear equation of the system. By summing up the different amplitudes of the previous sample's data and adding the estimation error, the amplitude of the signal at a given period is obtained in the AR method [14].

There are three kinds of solution of the AR model, the Yule–Walker method, Burg method, and Covariance method. Details have been published in [14]. Compared with the periodogram in Figure 7.4, an example of the Burg method for AR parameter estimation is shown in Figure 7.8. We can see that the spectral curves via the Burg method are obviously smoother than the periodogram method, which illustrates that their variance is less than the classical PSD estimate method. Moreover, the frequency resolution of Burg method is markedly higher than that of periodogram method.

To summarize, the spectrum estimation techniques can be categorized as nonparametric or parametric (Matlab includes a third one called subspace). The nonparametric methods include the periodogram, the Welch modified periodogram, and the Blackman–Tukey methods. All these methods have the advantage of possible implementation using the FFT, but their disadvantages are the short data lengths of limited frequency resolution. Also, considerable care has to be exercised to obtain meaningful results. On the other hand, parametric methods can provide high resolution in addition to being computationally efficient. However, it is necessary to form a sufficiently

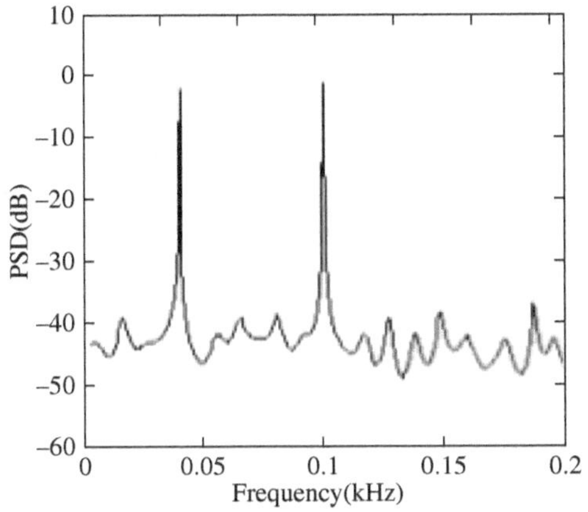

Figure 7.8 PSD estimation using the Burg method

accurate model of the process from which to estimate the spectrum. The most common parametric approach is to derive the spectrum from the parameters of an autoregressive model of the signal.

7.8 Conclusions

Power spectral analysis aims to describe the distribution (over frequency) of the power contained in a signal based on a continuous signal or a finite set of data. In order to deal with spectrum in a proper scientific manner, we returned to the main theme of spectral analysis. We discussed this topic with Fourier transform first and followed the concept of PSD. Then, the classic and modern PSD estimation methods were introduced.

References

[1] Leonowicz, Z., Lobos, T. and Rezmer, J. (2003) Advanced spectrum estimation methods for signal analysis in power electronics. *IEEE Trans. Industrial Electronics*, **50**(3), 514–519.
[2] Carbone, R., Menniti, D., Sorrentino, N., and Testa, A. (1998) Iterative harmonics and interharmonic analysis in multiconverter industrial systems. 8th International Conference on Harmonics and Quality of Power, Athens (Greece), pp. 432–438.
[3] Harris, B. (1967) *Spectral Analysis and Time Series*, Academic Press.
[4] Bendat, J.S. and Piersol, A.G. (2000) *Random Data Analysis and Measurement Procedures*, 3nd edn, John Wiley & Sons, Inc., New York.
[5] Meunier, M., Gif-sur-Yvette S., and Brouaye, F. (1998) Fourier transform, wavelets, prony analysis: tools for harmonics and quality of power. Proceeding of the 8th International Conference on Harmonics and Quality of Power, Athens (Greece), pp. 71–76.
[6] Priestley, M.B. (1982) *Spectral Analysis and Time Series*, Academic Press, London.
[7] Oppenheim, A.V., Schafer, R.W. and Buck, J.R. (1999) *Discrete-Time Signal Processing*, 2nd edn, Prentice-Hall, Upper Saddle River, NJ.
[8] Bartlett, M.S. (1948) Smoothing periodograms from time-series with continuous spectra. *Nature*, **161**, 686–687.

[9] Bartlett, M.S. (1950) Periodogram analysis and continuous spectra. *Biometrika*, **37**(1-2), 1–16.
[10] Welch, P.D. (1967) The use of fast Fourier Transform for the estimation of power spectra: a method based on time averaging over short, modified periodograms. *IEEE Trans. Audio Electroacoustics*, **15**(2), 70–73.
[11] Marple, S.L. Jr. (1989) A tutorial overview of modern spectral estimation. Proceeding of the International Conference on Acoustics, Speech, and Signal, 4, Glasgow, UK pp. 2152–2157.
[12] Blackman, R.B. and Tukey, J.W. (1959) *The Measurement of Power Spectra: from the Point of View of Communications Engineering*, Dover Publications, New York.
[13] Kay, S.M. (1988) *Modern Spectral Estimation: Theory and Application*, Prentice-Hall, Englewood Cliffs, NJ.
[14] Yilmaz, A.S., Alkan, A. and Asyali, M.H. (2008) Applications of parametric spectral estimation methods on detection of power system harmonics. *Electric Power Syst. Res.*, **78**(4), 683–693.

8

Dynamic Chaos Spectrum of Chaotic Switching Converters I: Wavelet Method

8.1 Introduction

For many signals, Fourier analysis is extremely useful because the signal's frequency contents are very important. So, why do we need other spectrum analysis techniques, like wavelet, prony analysis, or other methods? Actually, people have recognized for a long time that the Fourier transform has serious drawbacks, and these deficiencies are mainly reflected in three cases.

8.1.1 Lack of Time and Frequency Positioning

The time and frequency positioning function is defined as follows. For a given signal $x(t)$, we want to know which frequencies are involved at a specific moment or in a very short period of time. On the contrary, for a specific frequency or a very narrow frequency band, we also want to know when these frequency contents emerge. Looking at a Fourier transform, it is impossible to distinguish when a particular event took place. This is because time information is lost in transforming to the frequency domain by Equation 7.1 which is an average representation of the frequency characteristics throughout the time range in the whole integration interval. So, the Fourier transform lacks time and frequency positioning information.

8.1.2 Limitation for the Time-Variant Signals

If the signal properties do not change dramatically with time (that is to say, they are time invariant and called a stationary signal as mentioned previously), Fourier analysis

Chaos Analysis and Chaotic EMI Suppression of DC-DC Converters, First Edition. Bo Zhang and Xuemei Wang.

works well. However, most signals contain numerous nonstationary or transitory characteristics such as drift, trend, and abrupt change. The Fourier transform has limitations because any abrupt change in time of a nonstationary signal is spread out over the whole frequency axis. These characteristics are often the most important parts of the signals. For frequency analysis of nonstationary signals, we must adopt a statistical viewpoint, which deals with the characteristics of these signals.

8.1.3 Limitation for Resolution

Resolution, concluding time resolution and frequency resolution, indicates that the minimum interval may be figured out. Obviously, it will be a good spectral analysis if time resolution and frequency resolution may be automatically regulated to fit the signals' characteristics. Classic spectrum estimation, such as periodogram and BT methods based on the Fourier transform, suffers from the major problem of resolution, because time resolution and frequency resolution cannot be infinitely improved simultaneously based on the uncertainty principle [1]. In other words, a wave may be identified more accurately in frequency but be more blurred in space positions, and vice versa. There are inherent contradictions between time and frequency resolution in the Fourier transform.

These drawbacks limit the application of the Fourier transform in analyzing short sampled data sequences. The short sampled data sequences occur frequently in practice because many process measurements are short in duration or have slowly time-varying spectrums that are often considered as constant in short sampling intervals. Further, the Fourier transform cannot efficiently estimate inner-harmonics, since it assumes the harmonics to be integral multiples of the fundamental frequency. Attempting to figure out the limitations of the fast Fourier transform (FFT), many alternative spectral estimation methods have been proposed over the last three to four decades.

For the spectrum analysis of the nonlinear behavior of switching converters, the Fourier transform's drawbacks are reflected as the following.

1. The most prominent drawback is the frequency resolution. Frequency resolution in hertz is roughly the reciprocal of the time interval in seconds over which sampled data are available [2]. In the case of chaotic pulse width modulation (PWM), switching frequency varies quickly from time to time in microseconds, and the power spectral density on FFT reflects this nature with limited resolution.
2. The second is due to the limitations of data windows. In the case of chaotic PWM, converters present a stochastic nature, and there are a lot of harmonics and inter-harmonics. The connotative assumption of FFT is that harmonics have integral multiples of the fundamental frequency, so FFT is only suitable for the stationary and periodical signals [3–6].

A straightforward approach to overcome this problem is to perform the Wavelet Transform (WT). WT can track down transient electrical power systems with precision. Compared with FFT, another benefit of WT is that it is a powerful tool for time–frequency behavior of finite-energy signals [7–11].

The main source of electromagnetic interference (EMI) emissions in power converters comes from the switching of DC voltages following a constant scheme. Waveforms of interest are typically periodic functions of time. The peaks of EMI are mainly concentrated on multiples of switching frequency. In the case of chaotic PWM, even though the energy of harmonics is spread over the frequency belt, the peak of harmonics greatly decreases. It is meaningful to study the relationships between PWM schemes and conduct EMI of the system based on time–frequency and WT [12–22]. This chapter parameterizes time-domain power excursions that are important to both peak EMI emission constraints and component stressing (which the Fourier approach does not do as well).

8.2 Basic Principle of Wavelet Analysis

The Fourier transform is a tool widely used for many scientific purposes, but it is well suited only to the study of stationary signals where all frequencies have an infinite coherence time. The Fourier analysis brings only global information which is not sufficient to detect compact patterns. A straightforward solution to overcome the limitations of the Fourier transform is to introduce an analysis window of certain length that glides through the signal along the time axis to perform a "time-localized" Fourier transform. Such a concept led to the short-time Fourier transform (STFT), introduced by Dennis Gabor in his paper titled "Theory of communication," published in 1946 [23]. The STFT employs a sliding window function. For each specific t, a time-localized Fourier transform is performed on the signal $x(t)$ within the window. Through such operations, a Fourier transform of the entire signal can be performed. The signal segment within the gliding window is assumed to be approximately stationary. Unfortunately, this transform also has shortcomings, mostly in that it cannot get decent resolutions for both high and low frequencies at the same time. The Wavelet is a new transform which is generally much more concentrated in time. Moreover, it usually provides an analysis of the signal which is localized in both time and frequency, while the Fourier transform is localized only in frequency.

A wavelet is a wave-like oscillation with the amplitude, which increases from zero and then decreases back to zero. There are three examples of mother wavelets: Meyer, Morlet, and Mexican hat. The Morlet complex wavelet has the properties of high resolution and low aliasing as plotted in Figure 8.1 [23]. Unlike the sine wave used in the Fourier transform, which decomposes a signal, wavelets are generally much more concentrated in time [24]. WT is the local analysis of time (spatial) frequency by scaling and translating the signal, eventually achieving the goal of multiscale.

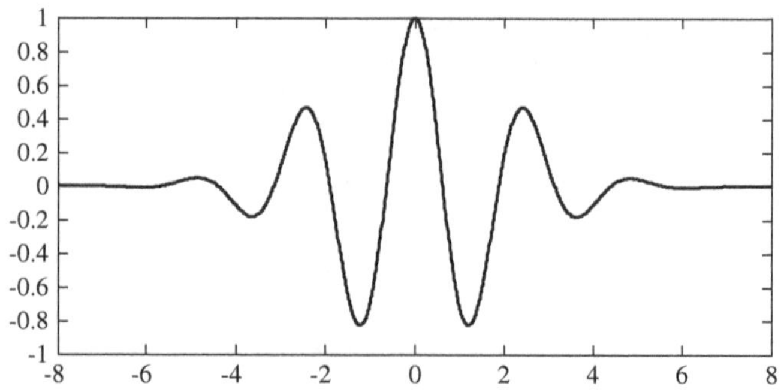

Figure 8.1 Morlet wavelet [23]

Generally, the WT can be represented in continuous (i.e., Continuous Wavelet Transform (CWT)) as well as in discrete forms (i.e., Discrete Wavelet Transform (DWT)). Given a mother wavelet $\psi(t)$ (which can be considered simply as a basic function in L^2, a quadratic integrable function), the CWT of a function $x(t)$ (assuming $x(t) \in L^2(R)$) is defined as [25]:

$$WT(a,b) = \frac{1}{\sqrt{a}} \int_{-\infty}^{\infty} x(t)\psi^* \left(\frac{t-b}{a}\right) dt \quad a \neq 0 \tag{8.1}$$

or as inner product form

$$WT_f(a,b) = \langle f, \psi_{a,b} \rangle \tag{8.2}$$

where ψ^*(t) is the complex conjugate of wavelet function ψ(t), $\psi_{a,b}(t) = |a|^{-1/2}\psi\left(\frac{t-b}{a}\right)$ and the other parameters are explained as follows. The scaling or dilation parameter a corresponds to frequency information ($a \propto$frequency^{-1}). The shift (translation along the time axis) parameter $\boldsymbol{b}$ relates to the location of the wavelet function, so it corresponds to the time information in the transform. $1/\sqrt{a}$ is an energy normalization factor. These two wavelet parameters transforming a signal with the wavelet basis mean that such a signal will be projected into a 2D, time-scale plane, instead of the 1D frequency domain in the Fourier transform [24]. This transform process is illustrated in Figure 8.2 [26]. We can understand the meaning of the WT from the following example. When we observe a target $x(t)$ using a lens, $\psi(t)$ just plays the role of a lens, b plays the role of shifting the lens to aim the target, and a plays the role of the tracking shot.

Now, some simple conclusions may be summarized. The name wavelet implies that it is a "small wave" of a waveform of effectively limited duration with zero mean. "Small" refers to attenuation, and "wave" refers to oscillation of amplitude. Compared with the Fourier transform decomposing the signal into a series of different frequencies of the sine wave, wavelet analysis decomposes the signal into a series of superpositions and all these wavelet functions come from shifting and scaling the mother wavelet function.

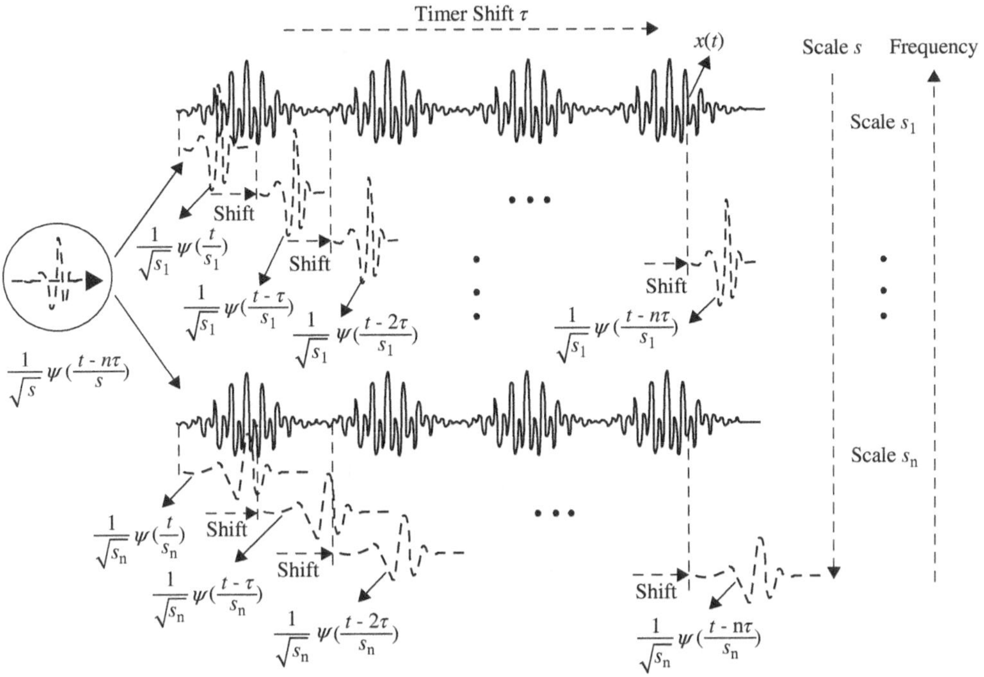

Figure 8.2 Illustration of wavelet transform [25]. With kind permission from Springer Science+Business Media: Springer Science, "Wavelets: Theory and Applications for Manufacturing", © 2011, Gao, R.X., Yan, R.

Calculating wavelet coefficients at every possible scale is a fair amount of work, and it generates an awful lot of data. What if we choose only a subset of scales and positions at which to make our calculations?

It turns out, rather remarkably, that if we choose scales and positions based on powers, then our analysis will be much more efficient and just as accurate. We obtain such an analysis from the DWT. For more information on DWT, see [27].

An efficient way to implement this scheme using filters was developed in 1988 by Mallat [28]. The Mallat algorithm is in fact a classical scheme known in the signal processing community as a two-channel subband coder.

Next, we briefly review the history of wavelet. The first recorded mention of what we now call a "wavelet" seems to be in 1909, in a dissertation titled "On the theory of the orthogonal function systems" by Alfred Haar. The concept of wavelets in its present theoretical form was first proposed by Jean Morlet to store and represent the seismic signal in oil exploration in 1984. The methods of wavelet analysis were developed mainly by Y. Meyer [29] and his colleagues, who ensured the methods' dissemination. The main algorithm dates back to the work of Stephane Mallat in 1988 [28]. Since then, the wavelet domain has grown very quickly. Many mathematical papers and practical trials were published every month. The research on wavelets is spearheaded by the work of scientists such as Ingrid Daubechies ("Ten Lectures on Wavelets" [30], played an important role in promoting studies of wavelet), Barbara Burke Hubbard (describes the birth, the history, and the seminal concepts in a very clear text [31]), Ronald Coifman, and Victor Wickerhauser. For more details about the theory and the properties of the WT, the tutorials in [32, 33] are strongly advised.

8.3 Multiresolution Analysis and Orthogonal Wavelets Basis

Multiresolution analysis (MRA) or multiscale approximation (MSA), introduced in Stephane Mallat [28, 34], are common methods used to analyze the DWT or verify the fast WT theory.

Multiresolution analysis is defined as the process of representing a signal by a finite sum of components at different resolutions such that each component can be processed adaptively based on the objectives of the application. It is the technique of representing signals compactly and at several levels of resolution to use for decomposition and reconstruction purposes.

For example, for a rapidly changing signal, what we are most interested in is the moment of occurrence of the change, and we only want to have a rough understanding for the frequency contents. On the other hand, for the slow changing parts of the signal, what we are interested in is their frequency contents, for example, whether it changes every second or every minute. Based on the uncertainty principle, high frequency resolution and high time resolution are not allowed simultaneously. So an ideal signal analysis method for this kind of signal is shown as the following. For the high frequency components, a window with high time resolution and low frequency resolution is adopted. And on the contrary, for the low frequency components, a window with high frequency resolution and low time resolution is adopted. Therefore, the concept of MRA was born to analyze the signal upon different resolutions.

A MRA of the space $L^2(R)$ consists of a sequence $\{V_j\}_{j\in Z}$ of nested subspaces

$$\{0\} \cdots \subset V_0 \subset V_1 \subset \cdots \subset V_n \subset V_{n+1} \subset \cdots L^2(R)$$

that satisfies certain self-similarity relations in time/space and scale/frequency, as well as completeness and regularity relations. Multiresolution has the following properties [35].

1. *Monotonicity*: for arbitrary $j \in Z$, there is $V_j \subset V_{j-1}$
2. *Denseness and completeness*: $\bigcap_{j\in Z} V_j = \{\varnothing\}$ and $\bigcup_{j=-\infty}^{\infty} V_j = L^2(R)$
3. *Scalability*: $\phi(t) \in V_j \Rightarrow \phi(2t) \in V_{j-1}$
4. *Translational invariance*: $\phi_j(t) \in V_j \Rightarrow \phi_j(t - 2^{j-1}k) \in V_j \qquad \forall k \in Z$
5. *Beingness of Riesz basis*: there is $\phi(t) \in V_0$, which made $\{\phi(2^{-j}t - k)\}_{k\in Z}$ constructing Riesz basis of V_{j}.

The MRA is based on the scaling function $\phi(t)$. The scaling function is a continuous, square integrable and, in general, real-valued function. The scaling function does not satisfy the wavelet admission condition: the mean value of the scaling function $\phi(t)$ is not equal to zero, but is usually normalized to unity.

The basic scaling function $\phi(t)$ is dilated by dyadic scale factors k (scaling factor) and j (translation factor). At each scale level the scaling function is shifted by discrete translation factors as

$$\phi_{j,k}(t) = 2^{\frac{-j}{2}}\phi(2^j t - k) \quad j, k \in Z,\ t \in R \tag{8.3}$$

The scaling functions of all scales 2^j with $j \in Z$ generated from the same $\phi(t)$ are all similar in shape. At each resolution level j, the set of the discrete translations of the scaling functions, $\phi_{j,k}(t)$, forms a function basis that spans a subspace

$$V_j = \overline{span\left\{\phi_{j,k}(t)\right\}} \tag{8.4}$$

This means that if a arbitrary function $f(t) \in V_j$, it can be expressed as

$$f(t) = \sum_k \alpha_k \phi_{j,k}(t) = 2^{\frac{-j}{2}} \sum_k \alpha_k \phi(2^{-j}t - k) \tag{8.5}$$

If j increases, the definition domain of $\phi_{j,k}(t)$ becomes larger, and the spanned space by linear combination Equation 8.5 only includes large-scale slowly varying signals. Whereas Equation 8.5 may represent the function's more subtle, slight changes. According to $L^2(R) = \bigcup_{j=-\infty}^{\infty} V_j$ and $V_{j-1} = V_j \oplus W_j$, we have

$$L^2(R) = \bigoplus_{j=-\infty}^{j=\infty} W_j \tag{8.6}$$

where W_j is the complementary space of V_j in V_{j-1}, and the orthogonal wavelet bases are

$$\psi_{jk}(t) = 2^{\frac{-j}{2}} \sum_k \alpha_k \psi(2^{-j}t - k) \tag{8.7}$$

where ψ is the wavelet function and W_j is the wavelet space with scale j. The wavelet space is the difference of adjacent scale spaces. $W_j = V_{j-1} - V_j$ represents the small differences of projections between adjacent scale spaces, which is the projection of $f(t)$ on wavelet space and also calls detail space.

Orthogonal wavelet bases satisfy the following properties

$$\int \psi_{jk}^p(t)\psi_{jk}^q(t)dt = \begin{cases} 0 & \text{if } p \neq q \\ 1 & \text{if } p = q \end{cases}$$

$$\int (\psi_{jk}(t))^2 dt = 1$$

$$\int \phi_{jk}(t)\psi_{jk}^p(t)dt = 1 \quad p \neq 0$$

Approximation $P_{V_j} f$ of function $f(t)$ on the resolution 2^{-j} is defined as its orthogonality projection on V_j

$$P_{V_j} f = \sum_{k=-\infty}^{\infty} < f \quad \phi_{jk} > \phi_{jk} \tag{8.8}$$

The inner product is

$$\alpha_j[k] = < f, \phi_{jk} > = \int_{-\infty}^{\infty} f(t) 2^{\frac{-j}{2}} \phi(2^j t - k) dt = f^* \overline{\phi_j}(2^j k) \tag{8.9}$$

it provides a discrete approximation of $f(t)$ on the resolution 2^{-j}.

8.4 Wavelet Transform and Filter Bank

The fast WT continuously decomposes each scale approximation $P_{V_j} f$ into the sum of scaling coefficients $P_{V_{j+1}} f$ on smaller scales. Because $\{\phi_{jk}\}_{k \in Z}$ and $\{\psi_{jk}\}_{k \in Z}$ are the orthonormal basis of V_j and W_j, the two spaces' projections may be described as

$$\alpha_j[k] = < f, \phi_{jk} > \tag{8.10}$$

and

$$d_j[k] = < f, \psi_{jk} > \tag{8.11}$$

These coefficients can be obtained by iterative calculations of discrete convolution and subsampling as

$$\alpha_{j+1}[k] = \sum_{n=-\infty}^{\infty} h[n - 2k] \alpha_j[n] = \alpha_j * \overline{h}[2k] \tag{8.12}$$

$$d_{j+1}[k] = \sum_{n=-\infty}^{\infty} g[n - 2k] \alpha_j[n] = \alpha_j * \overline{g}[2k] \tag{8.13}$$

The two equations indicate that d_{j+1} and a_{j+1} are obtained by convoluting a_j with $\overline{g}$ and $\overline{h}$ respectively followed by down-sampling in factor 2, and satisfying

$$g[k] = (-1)^n h[1 - k] \tag{8.14}$$

where h and g compose a pair of conjugate filters. The wavelet decomposing tree is demonstrated in Figure 8.3, where level 0 is the original signal.

8.5 Wavelet Analysis of Chaotic PWM

8.5.1 Basic Principle of Chaotic PWM Control

The basic principle of chaotic modulated PWM can be demonstrated as follows [36]. The carrier's frequency consists of two parts, one is the fixed switching frequency f_g,

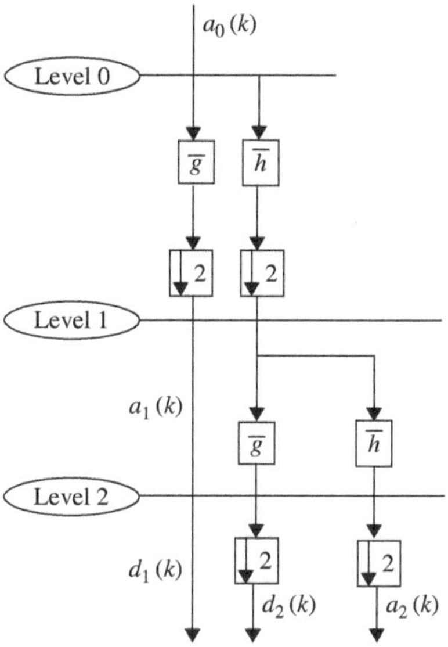

Figure 8.3 Wavelet decomposing tree

and another is the Δf_h generated by chaos generator and superimposed on carrier f_g. Namely, the chaotic frequency f can be expressed as

$$f = f_g + \Delta f_h \tag{8.15}$$

Obviously, when $\Delta f_h = 0$, it is fixed frequency PWM. When Δf_h changes periodically, it is periodic PWM such as sine modulated PWM. When Δf_h changes chaotically, it is chaos modulated PWM. Figures 8.4a–c illustrate the constant frequency PWM, sine modulated PWM, and chaos modulated PWM respectively.

The modulated PWM signal can be described as a driver signal $P(t)$, whose frequency is f and duty cycle is d, and serves as a signal source for wavelet and time–frequency analysis, and we have

$$P(t) = Square(f_g + \Delta f_h, d) \tag{8.16}$$

where t denotes time.

8.5.2 *Wavelet Analysis*

A complex Morlet wavelet has the properties of high resolution and low aliasing [34]. It is suitable for spectrum analysis in switching converters. A complex Morlet wavelet is defined as

$$\phi(t) = \exp(-t^2/2)e(i\omega_0 t) \tag{8.17}$$

$$\xi(\omega) = \sqrt{2\pi}\exp(-(\omega - \omega_0)^2/2) \tag{8.18}$$

where ω_0 is the wavelet center frequency.

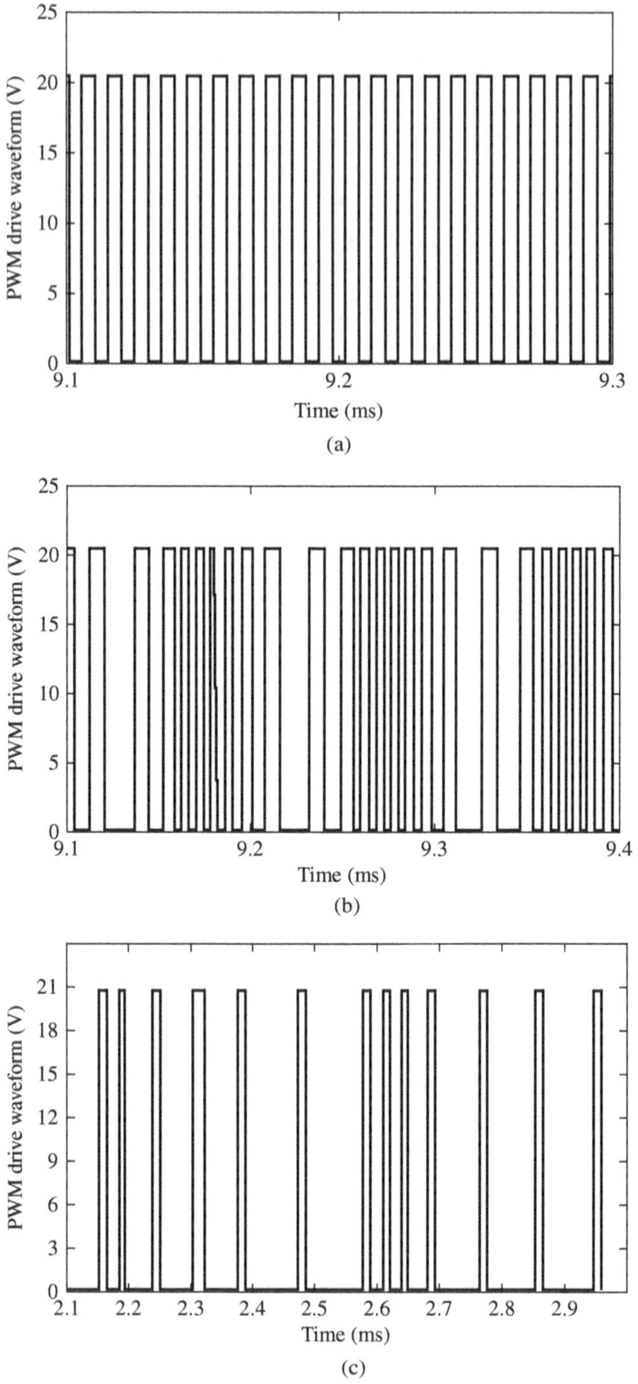

Figure 8.4 Waveform of PWM drive pulse with (a) constant PWM, (b) periodic PWM, and (c) chaotic PWM

Waveforms of a complex Morlet wavelet are illustrated in Figure 8.5. A complex Morlet wavelet has good characteristics of time–frequency analysis. When $\omega_0 \geq 5$, it meets admissibility conditions [9]. Using a complex Morlet wavelet, the continuous wavelet transformation of Equation 8.15 is obtained as follows

$$
\begin{aligned}
WT_P(\alpha, b) &= \langle P(t), \phi_{\alpha,b}(t)\rangle \\
&= \frac{1}{\alpha}\int_R P(t)\phi\left(\frac{t-b}{\alpha}\right)dt \\
&= \frac{1}{\alpha}\int_R P(t)\exp\left(-\frac{(t-b)^2}{2\alpha^2}\right)\exp\left(-i\frac{\omega_0}{\alpha}(t-b)\right)dt
\end{aligned}
\tag{8.19}
$$

The ordinate a represents the scale index, and the abscissa b indicates time shifting (or translation). The coefficients of a PWM drive pulse translated by a complex Morlet wavelet are illustrated in Figure 8.6. As the shade darkens, the coefficients are decreasing. Figures 8.6a,b illustrate the wavelet coefficients of constant PWM and periodic PWM, which are matched for similarities at the time domain. Figure 8.6c depicts that the wavelet coefficients of chaotic PWM have no regulation at the time axis, it is determined by the chaos characteristic of chaotic PWM.

8.5.3 Wavelet Reconstruction of Chaotic PWM

In order to further analyze the characteristics of PWM drive pulse $P(t)$, Equation 8.16 is decomposed and reconstructed [36]. $\phi(t)$ and $\psi(t)$ are the wavelet function and the scale function respectively, with the resolution of approaching 2^{-j}. The part of approaching $A_jP(t)$ and the part of detail $D_jP(t)$ can be derived

$$A_jP(t) = \sum_{K=-\infty}^{\infty} a_{j,k}\psi_{j,k}(t) \tag{8.20}$$

$$D_jP(t) = \sum_{K=-\infty}^{\infty} d_{j,k}\phi_{j,k}(t) \tag{8.21}$$

where $a_{j,k}$ is the coefficient of approaching and $d_{j,k}$ is the coefficient of detail with the resolution of approaching 2^{-j} . The Mallet algorithm [28] is used for the WT of PWM drive pulse $P(t)$

$$A_jP(t) = A_{j+1}P(t) + D_{j+1}P(t) \tag{8.22}$$

where

$$A_{j+1}P(t) = \sum_{m=-\infty}^{\infty} a_{j+1,m}\psi_{j+1,m}(t) \tag{8.23}$$

$$D_{j+1}P(t) = \sum_{m=-\infty}^{\infty} d_{j+1,m}\phi_{j+1,m}(t) \tag{8.24}$$

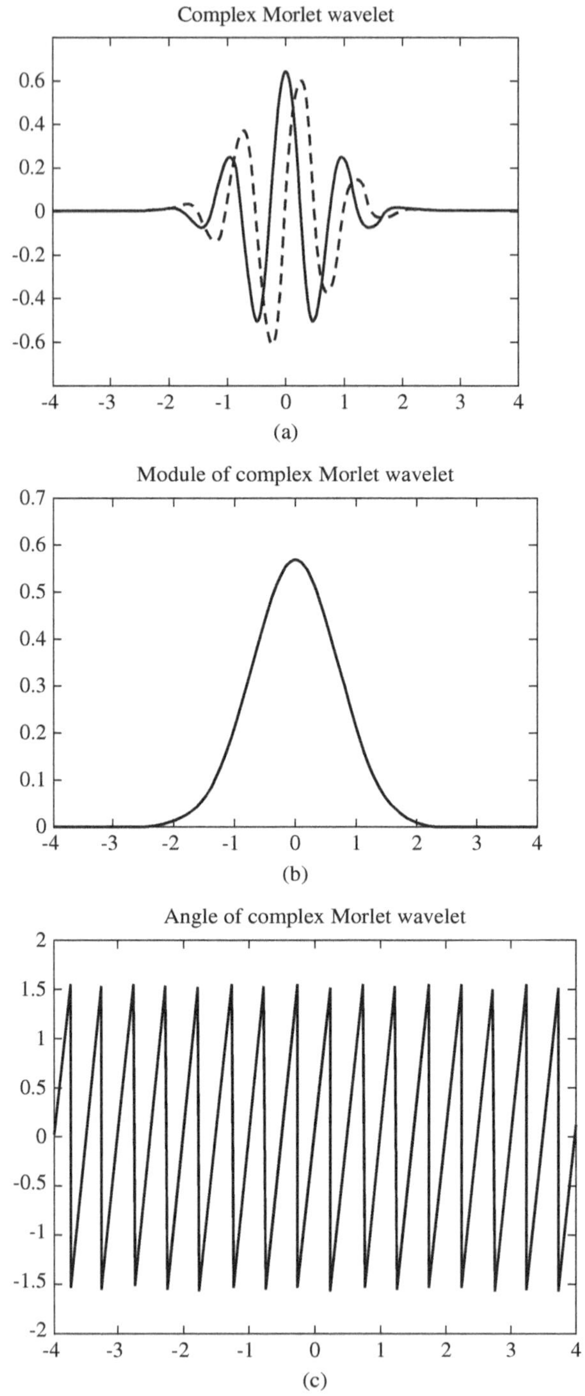

Figure 8.5 Complex Morlet wavelet for (a) time domain (the real line describes a real part, and the dashed line describes an imaginary part), (b) module, and (c) angle

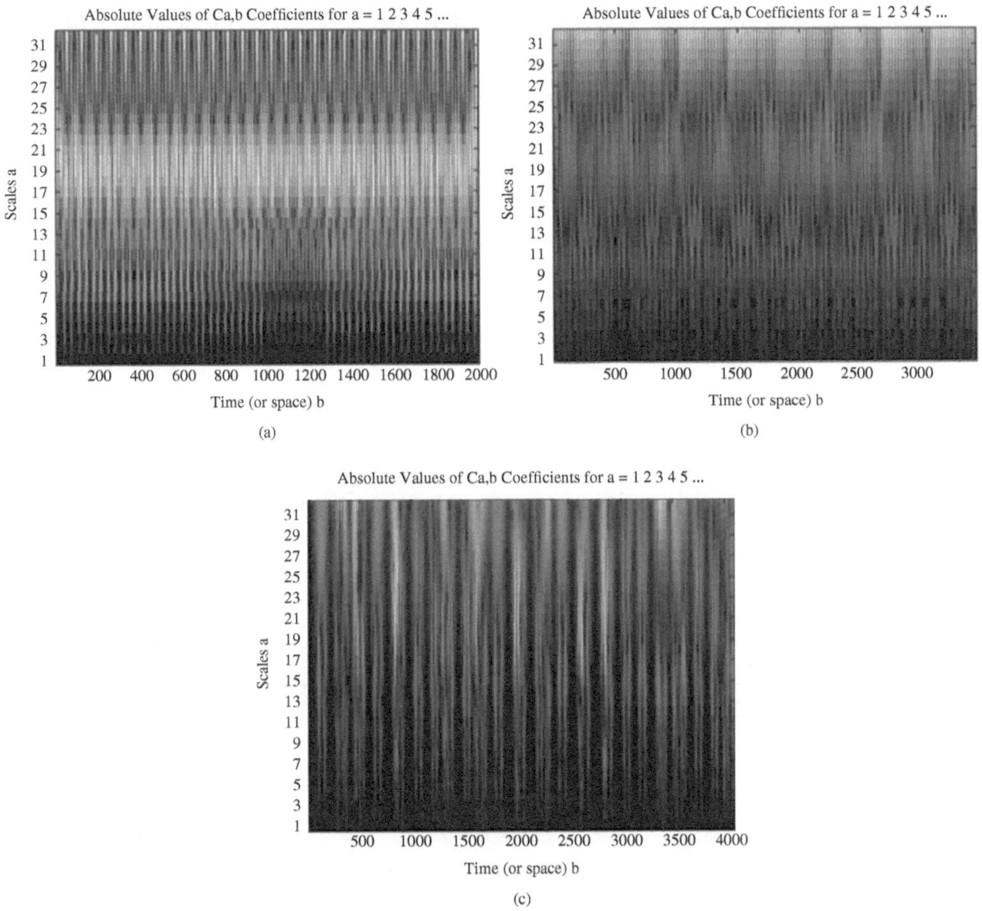

Figure 8.6 Wavelet coefficients of PWM signals with (a) constant PWM; (b) periodic PWM; and (c) chaotic PWM

Thus,

$$\sum_{m=-\infty}^{\infty} a_{j+1,m}\psi_{j+1,m}(t) + \sum_{m=-\infty}^{\infty} d_{j+1,m}\phi_{j+1,m}(t) = \sum_{m=-\infty}^{\infty} a_{j,k}\psi_{j,k}(t) \tag{8.25}$$

Because the wavelet function is an orthonormal wavelet, the scale function is an orthonormal basis. Hence,

$$\begin{aligned}\psi_{j+1,m}(t) &= 2^{-(j+1)/2}\psi(2^{-(j+1)/2}t - m)\\ &= 2^{-(j+1)/2}\sqrt{2}\sum_{i=-\infty}^{\infty} h(i)\,\psi(2^{-j}t - 2m - i)\end{aligned} \tag{8.26}$$

On multiplying both sides of Equation 8.26 by $\psi^*_{j,k}(x)$, and integrating with t, we have

$$\psi_{j+1,m}(t) = 2^{-(j+1)/2}\psi(2^{-(j+1)/2}t - m)$$
$$= 2^{-(j+1)/2} \cdot \sqrt{2}\sum_{i=-\infty}^{\infty} h(i)\,\psi(2^{-j}t - 2m - i) \tag{8.27}$$

$$\langle \psi_{j+1,m}, \psi_{j,k}\rangle = \sum_{k=-\infty}^{\infty} h^*(k - 2m) \tag{8.28}$$

Similarly, we can obtain

$$\langle \psi_{j,m}, \phi_{j+1,m}\rangle = \sum_{k=-\infty}^{\infty} g^*(k - 2m) \tag{8.29}$$

On multiplying both sides of Equation 8.20 by $\psi^*_{j,k}(x)$, integrating with t and using Equation 8.28, we have

$$a_{j+1,m} = \sum_{k=-\infty}^{\infty} h^*(k - 2m)a_{j,k} \tag{8.30}$$

On multiplying both sides of Equation 8.24 by $\psi^*_{j,k}(x)$, integrating with t and using Equation 8.29, we have

$$d_{j+1,m} = \sum_{k=-\infty}^{\infty} h^*(k - 2m)a_{j,k} \tag{8.31}$$

On multiplying both sides of Equation 8.26 by $\psi^*_{j,k}(x)$, integrating with t and using Equations 8.28 and 8.29, we have

$$a_{j,k} = \sum_{k=-\infty}^{\infty} h(k - 2m)a_{j+1,k} + \sum_{k=-\infty}^{\infty} g(k - 2m)d_{j+1,k}$$

If the infinite matrix $H_{m,k} = h^*(k-2m)$ and $\mathrm{G}_{m,k} = g^*(k-2m)$ are introduced, we can rewrite Equations 8.30 and 8.31 as

$$\begin{cases} a_{j+1} = Ha_j \\ d_{j+1} = Ga_j \end{cases} \qquad j = 0, 1, 2 \cdots J \tag{8.32}$$

Thus, wavelet coefficients of the approximations and of the details are gained. Then, the reconstructed PWM drive pulse $P(\mathrm{t})$ also yields

$$a'_j = H^* a_{j+1} + G^* d_{j+1} \tag{8.33}$$

where H^* and G^* are dual operators of H and G respectively, the function of filter bank H is to approach the PWM drive pulse $P(t)$, and the function of filter bank G is to extract detailed information from the PWM drive pulse $P(t)$.

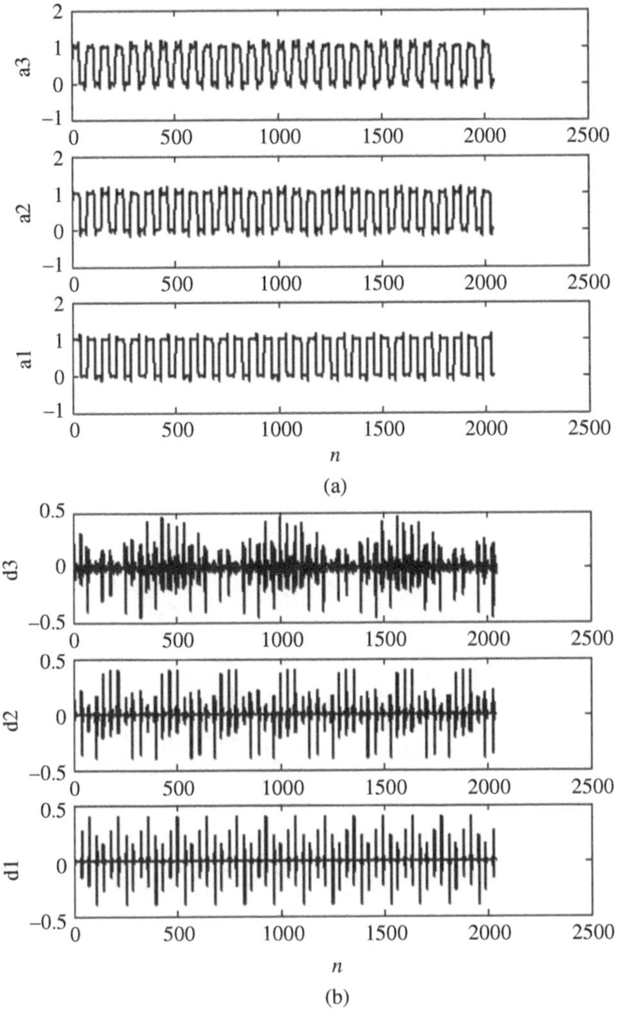

Figure 8.7 Reconstructed signals of constant PWM for (a) approaching signals of different layers and (b) detail signals of different layers

The approximations and details of different layers of the PWM drive pulse are depicted in Figures 8.7–8.9, which also show the responses of different layers to different scales. The figures reveal that the portion of low frequency (the signal of approaching) approaches the original signal, and the portion of high frequency contains the details or differences of the PWM scheme. The energy of the PWM mainly focuses on the low-frequency band.

The error list of the wavelet reconstruction is shown in Table 8.1. The error of the wavelet reconstruction is infinitesimal, and the precision of wavelet reconstruction is pretty high. Moreover, it is recommended that WT be employed to analyze a chaotic converter that has aneuploid inner harmonics.

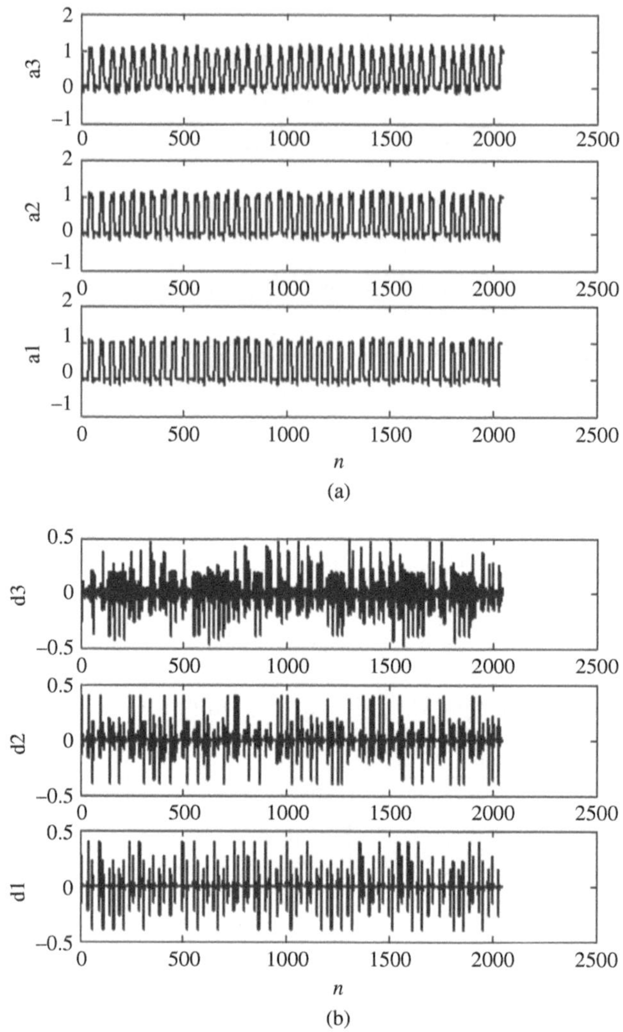

Figure 8.8 Reconstructed signals of periodic PWM for (a) approaching signals of different layers and (b) detail signals of different layers

Different scales a have different resolutions of time and frequency. The frequency characteristics of the signals can be detached by wavelet decomposition. Hilbert envelope and power spectrum analysis of different PWM schemes are accomplished as shown in Figures 8.10–8.12.

Compared with the constant PWM, the peak of the spectrum decreases about 15 dB in the case of periodic PWM and about 25 dB in the case of chaotic PWM. The energy of the signal spreads over the frequency domain, and the gross energy increases at the band of low frequency.

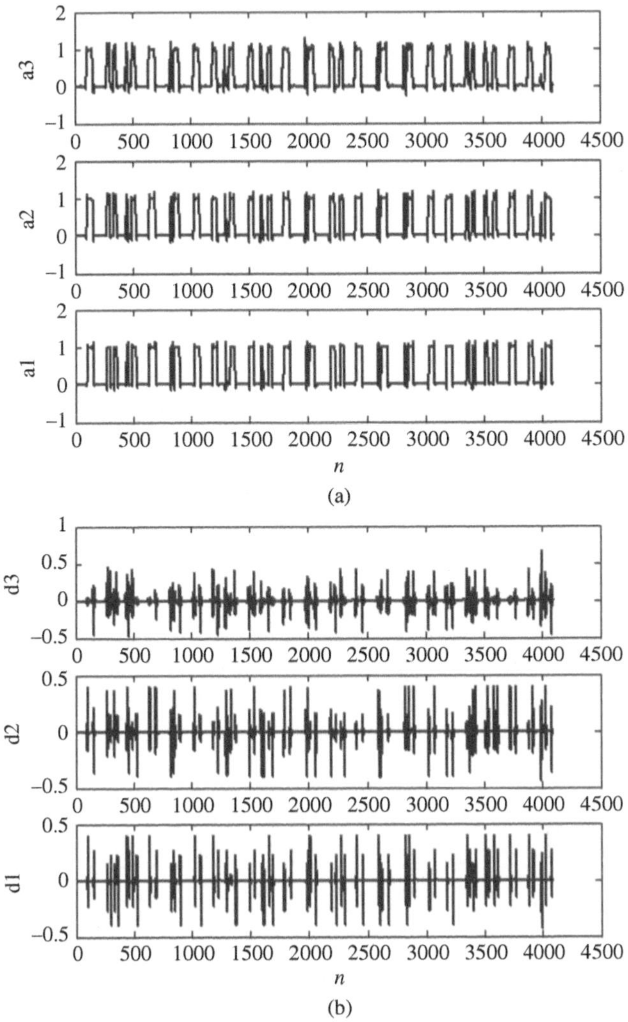

Figure 8.9 Reconstructed signals of chaotic PWM for (a) approaching signals of different layers and (b) detail signals of different layers

Table 8.1 Error list of wavelet reconstruction of $P(t)$

Constant PWM	Periodic PWM	Chaotic PWM
3.9931e-010	4.1976e-010	3.6546e-010

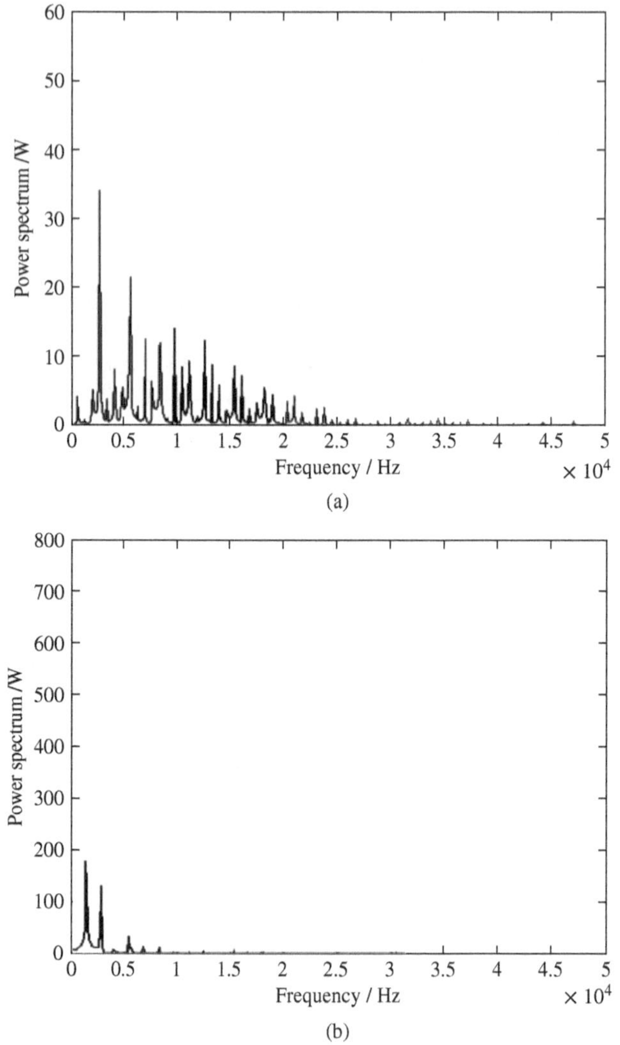

Figure 8.10 Hilbert envelope power spectrum of constant PWM for (a) detail signals of the first layer and (b) approaching signals of the third layer

8.5.4 Time-Frequency Analysis of the Chaotic PWM

The energy evolvement of the PWM drive pulse $P(t)$ is presented by a scalogram. The scalogram of the PWM drive pulse $P(t)$ is defined by the module of continuous WT

$$SC_P(\alpha, b) = |WT_P(\alpha, b)|^2 = \frac{1}{|\alpha|}\left|\int_{-\infty}^{\infty} P(t)\,\psi^*\left(\frac{t-b}{\alpha}\right)dt\right|^2 \tag{8.34}$$

It represents a local time–frequency energy density by measuring the energy of $P(t)$ in the Heisenberg box of each wavelet. This description of the energy of the signal

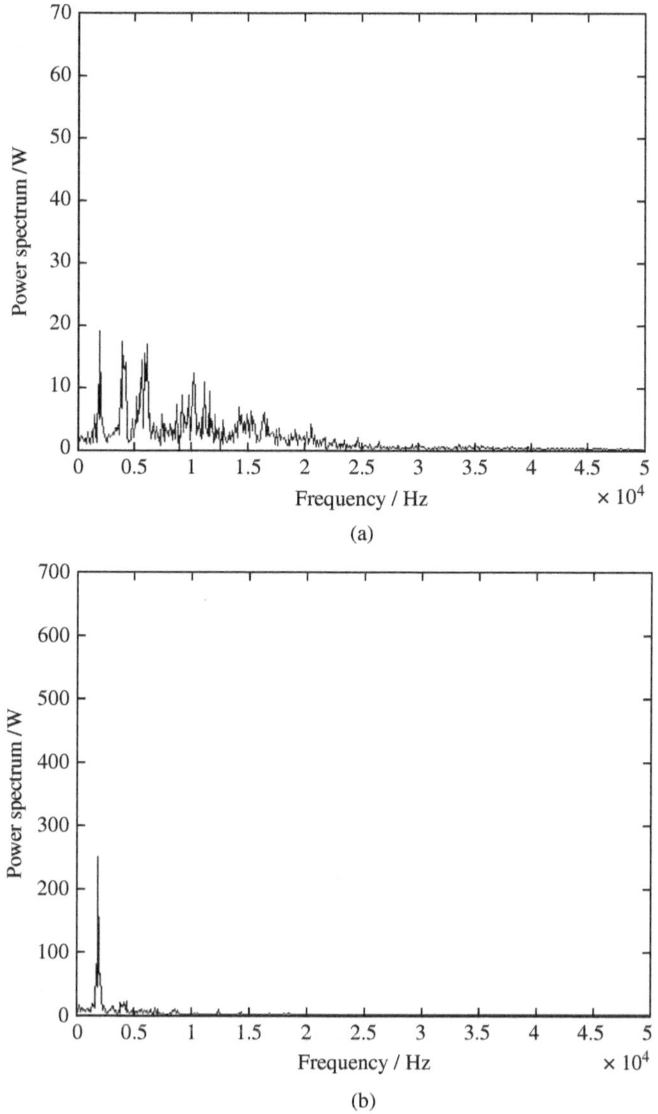

Figure 8.11 Hilbert envelope power spectrum of periodic PWM for (a) detail signals of the first layer and (b) approaching signals of the third layer

(in a timescale domain) facilitates identification of the time-varying energy flux. The spectral evolution and the transient bursts are not readily visible in the time- or the frequency-domain representations such as FFT.

The scalograms of the constant PWM, periodic PWM, and chaotic PWM are illustrated in Figures 8.13–8.15, respectively. The maximum energy ratio (MER) of $P(\mathrm{t})$ of the constant PWM, periodic PWM, and chaotic PWM are shown in Table 8.2.

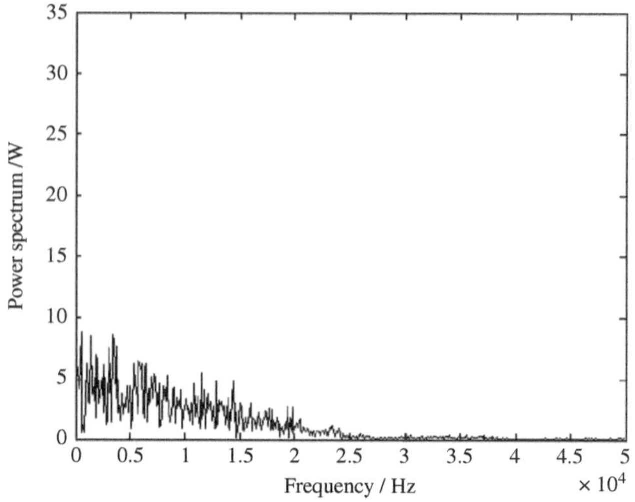

(a)

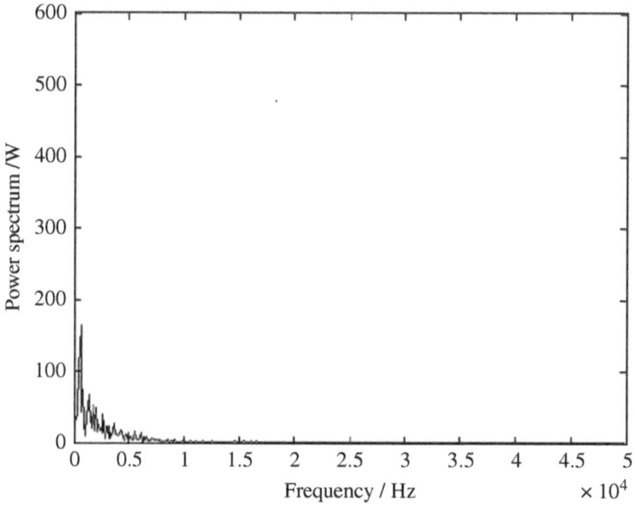

(b)

Figure 8.12 Hilbert envelope power spectrum of chaotic PWM for (a) detail signals of the first layer and (b) approaching signals of the third layer

The contour maps of scalogram distribution illustrate the relationship between time and frequency. In the case of constant PWM, the contour map is almost a beeline. In the case of periodic PWM, the contour map changes periodically. And in the case of chaotic PWM, the contour map has the characteristic of irregularity.

The 3D graphics of the scalogram illustrate the relationships among time, frequency, and energy. The energy of the system has a uniform distribution in constant PWM; the energy varies regularly in periodic PWM. In the case of chaotic PWM,

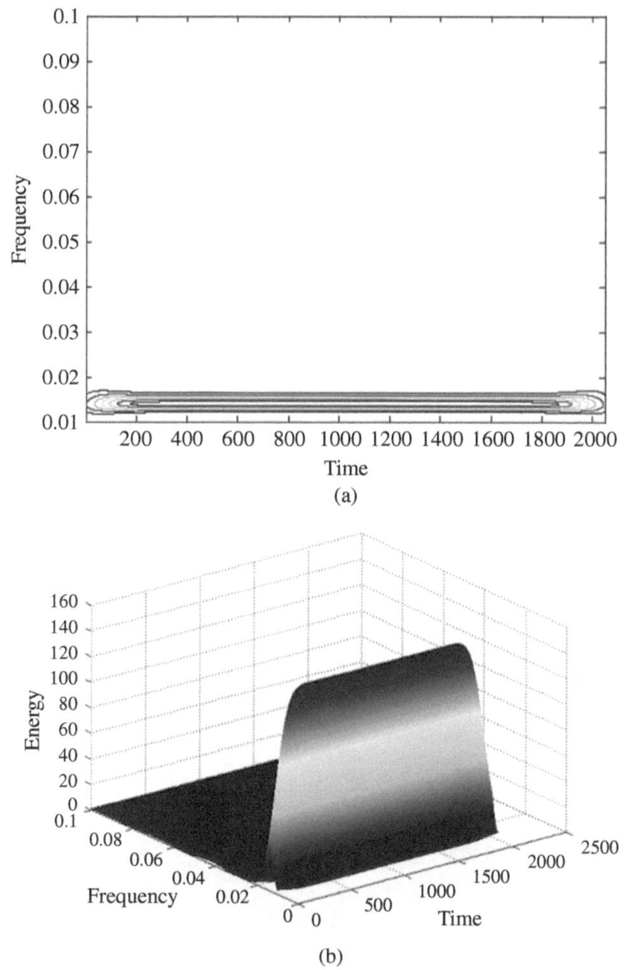

Figure 8.13 (a) Contour map and (b) 3D diagram of constant PWM

energy varies irregularly. Figures 8.13–8.15 reveal an important fact that the peak of energy decreases with the frequency spreading PWM scheme. Especially in chaotic PWM, energy spreads over the frequency domain.

The MER in decibels is decided by the scalogram

$$\mathrm{MER} = 20\mathrm{Log}_{10}\frac{\max_{\omega}(\mathrm{CON}(\omega, t))}{\max_{\omega}(\mathrm{SPWM}(\omega, t))} \tag{8.35}$$

where $\mathrm{CON}(\omega, t)$ is a scalogram of an unmodulated PWM signal and SPWM is a scalogram of a modulated switching signal. Table 8.2 gives an MER of the three PWM schemes. From the view of the time–frequency–energy distribution, the peak of energy has been cut down by chaotic PWM.

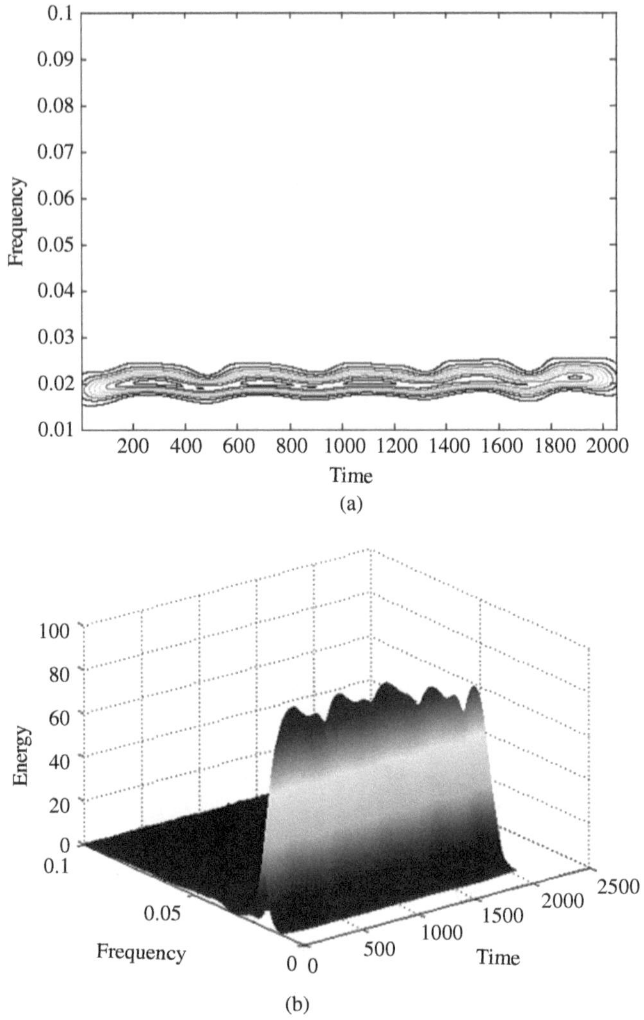

Figure 8.14 (a) Contour map and (b) 3D diagram of periodic PWM

8.5.5 Information on the Time–Frequency Image of P(t)

8.5.5.1 Renyi Information

One of the important parts of the information in the time–frequency transform is the number of essence components, namely the Renyi information. The third-order Renyi information of the time–frequency distribution $P_x(t,f)$ is

$$R_P^3 = -\frac{1}{2}\log_2\left\{\int_{-\infty}^{\infty}\int_{-\infty}^{\infty} P_x^{\ 3}(t,f)\,dtdf\right\} \tag{8.36}$$

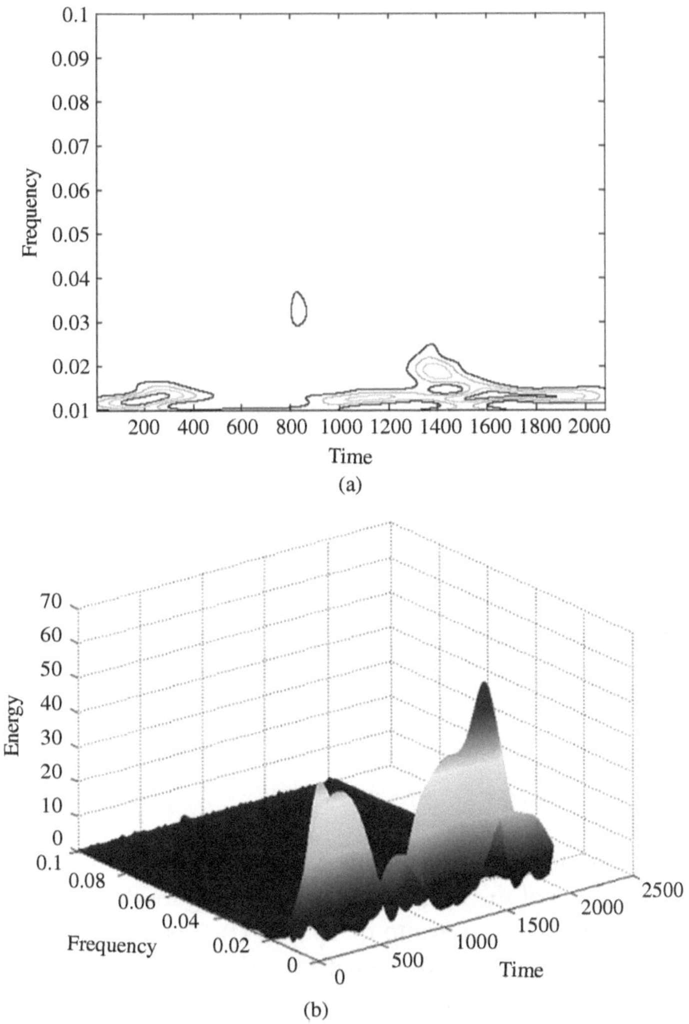

Figure 8.15 (a) Contour map and (b) 3D diagram of chaotic PWM

Table 8.2 MER of $P(t)$

Constant PWM	Periodic PWM	Chaotic PWM
1	1.9176	2.875

Table 8.3 Renyi information of PWM drive pulse *P(t)*

Constant PWM	Periodic PWM	Chaotic PWM
0.8936	1.1252	2.2229

Table 8.3 gives the Renyi information of the PWM drive pulse P(t) with different PWM schemes.

8.5.5.2 Torque Characteristics

The first-order time torque of time–frequency–energy distribution *tfr* of the PWM drive pulse is defined as

$$f_m(t) = \frac{\int_{-\infty}^{\infty} f tfr(t,f)df}{\int_{-\infty}^{\infty} tfr(t,f)df} \tag{8.37}$$

The first-order frequency torque is defined as

$$B^2(t) = \frac{\int_{-\infty}^{\infty} f^2 tfr(t,f)df}{\int_{-\infty}^{\infty} tfr(t,f)df} - f_m(t)^2 \tag{8.38}$$

The first-order time torque corresponds to the transient frequency of $P(t)$, and it possesses characteristics of the transient frequency (Figure 8.16). The first-order frequency torque corresponds to a group delay of $P(t)$ (Figure 8.17). The characteristics of torque are in perfect agreement with the drive pulse waveform as shown in Figure 8.4 and with the scalogram diagrams as depicted in Figures 8.13–8.15.

8.5.5.3 Marginal Characteristics

The time margin described by the distribution *tfr* of the PWM drive pulse $P(t)$ is

$$m_f(t) = \int_{-\infty}^{\infty} tfr(t,f)df \tag{8.39}$$

The frequency margin is defined as

$$m_t(f) = \int_{-\infty}^{\infty} tfr(t,f)dt \tag{8.40}$$

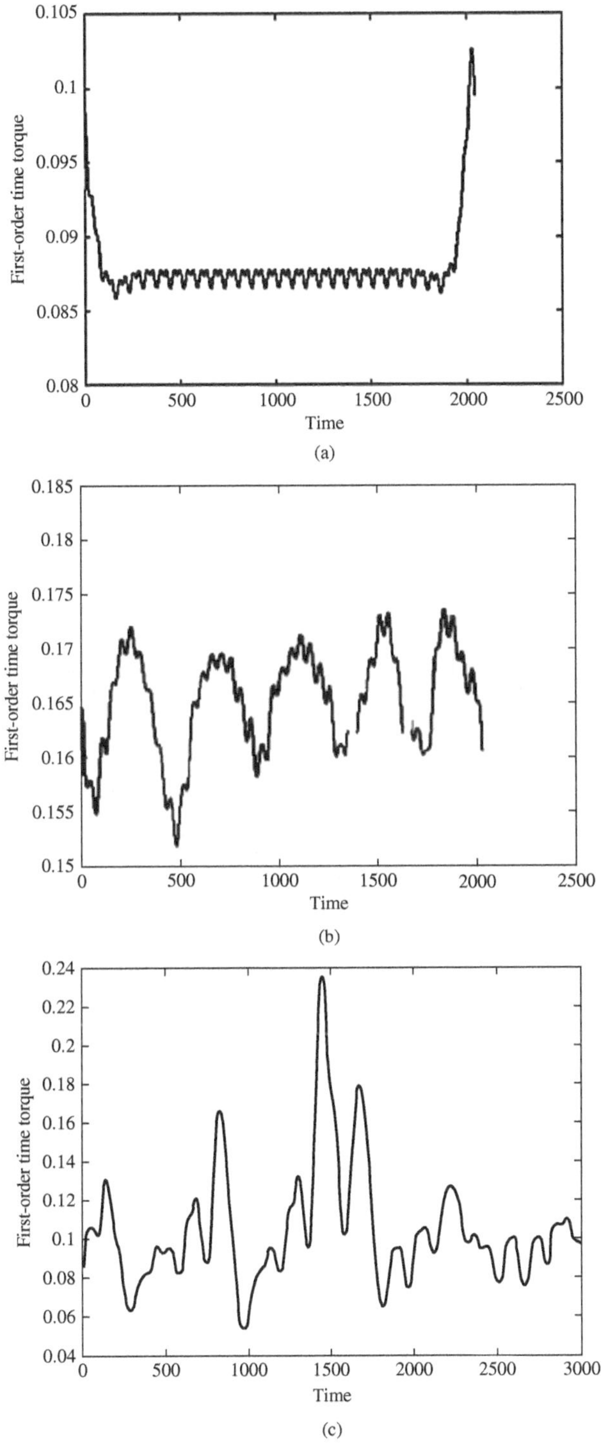

Figure 8.16 First-order time torque of $P(t)$ with (a) constant PWM, (b) periodic PWM, and (c) chaotic PWM

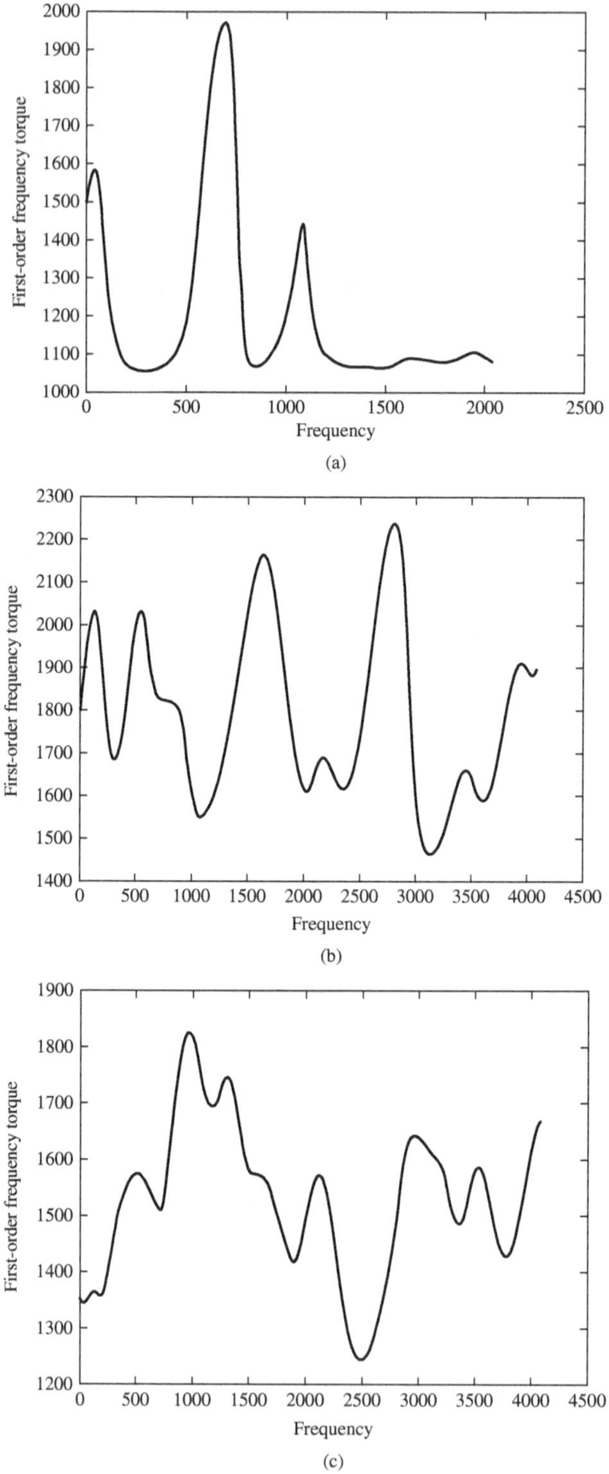

Figure 8.17 First-order frequency torque of P(t) with (a) constant PWM, (b) periodic PWM, and (c) chaotic PWM

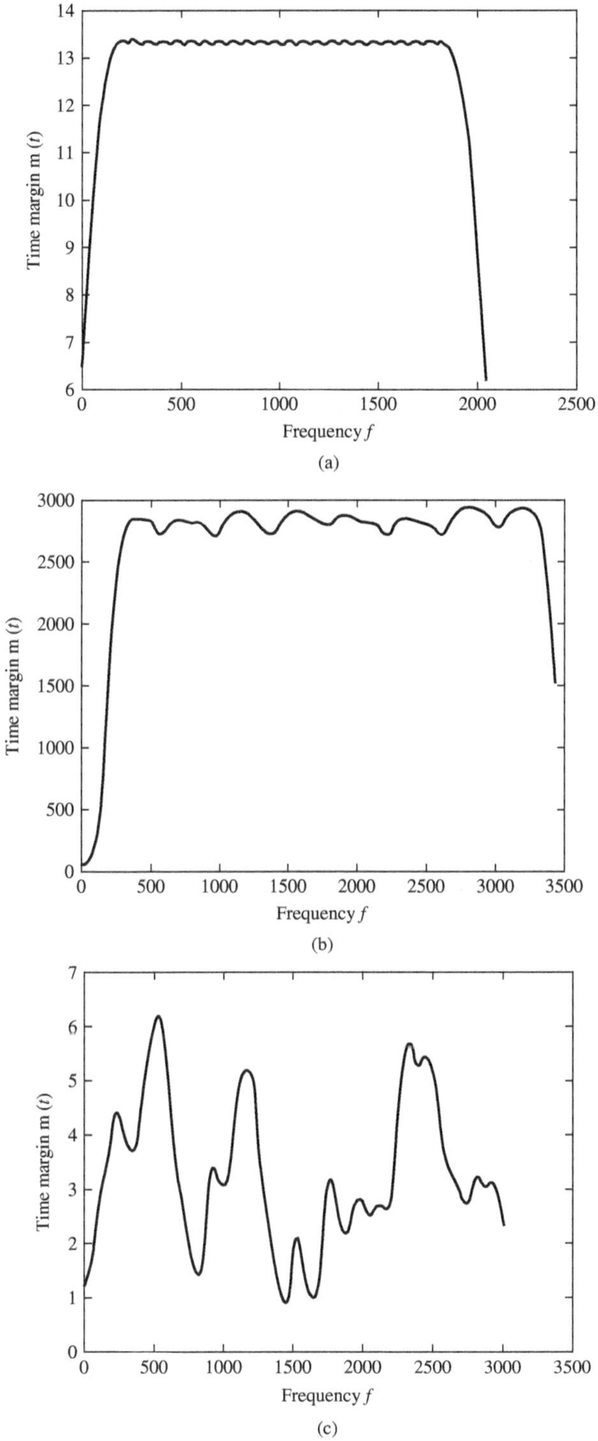

Figure 8.18 Time margin characteristics of P(t) with (a) constant PWM, (b) periodic PWM, and (c) chaotic PWM

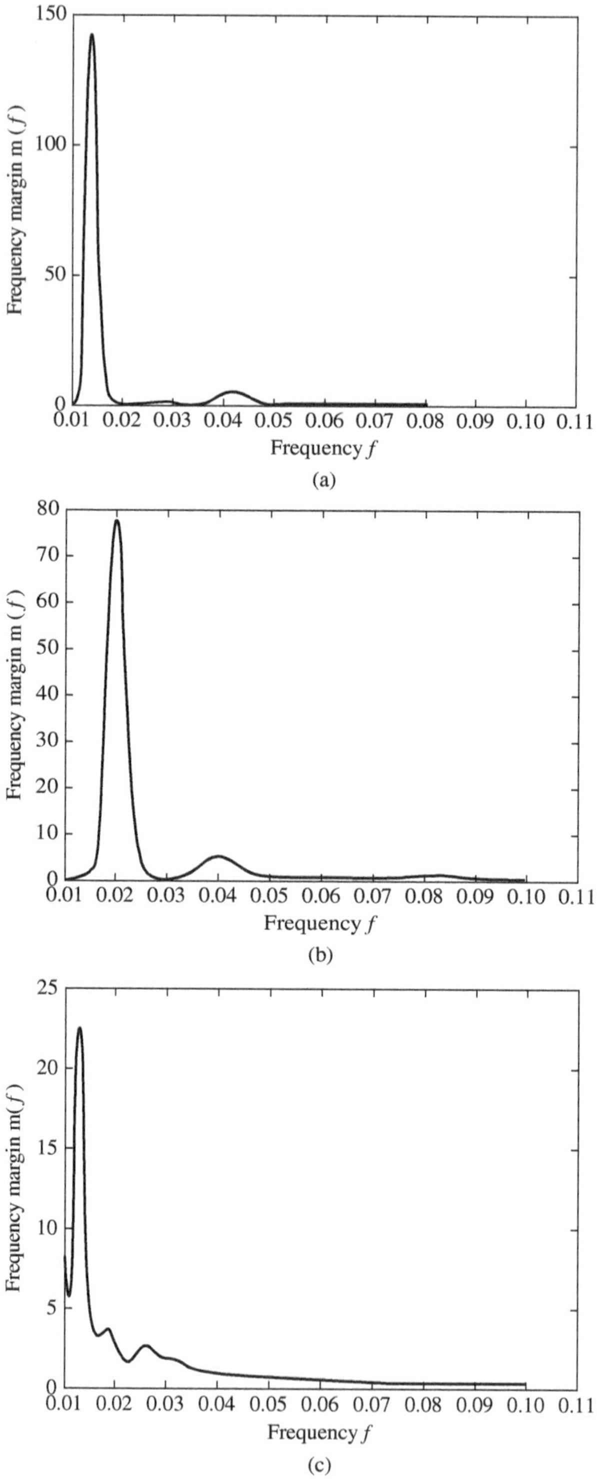

Figure 8.19 Frequency margin characteristics of PWM drive P(t) with (a) constant PWM, (b) periodic PWM, and (c) chaotic PWM

Table 8.4 Low frequency energy spectrum density of $P(t)$

Constant PWM	Periodic PWM	Chaotic PWM
145.1	78.8	22.7

In the physical sense, the time margin of the PWM drive pulse $P(t)$ corresponds to the transient energy, and the frequency margin of the PWM drive pulse $P(t)$ corresponds to the power spectrum density, that is

$$m_f(t) = |P(t)|^2 \tag{8.41}$$

$$m_t(f) = |P(f)|^2 \tag{8.42}$$

The time margins (see Figure 8.18) of the constant PWM reveal that transient energy has a uniform distribution at the time axis. The transient energy of the periodic PWM varies periodically. And the transient energy of the chaotic PWM varies irregularly. This result is in accordance with the scalogram diagram (see Figures 8.13–8.15).

The frequency margin of the PWM drive pulse $P(t)$ (see Figure 8.19) depicts that some energy of the middle frequency band shifts to the low frequency band in the case of chaotic PWM. And the energy spectrum density's peak of the low frequency band decreases after chaotic spectrum spreading, shown in Table 8.4.

8.6 Conclusions

The wavelet is a small wave that has an oscillating wavelike characteristic and has its energy concentrated in time. This chapter has given time–frequency–energy distributions and WTs of the three PWM schemes. Energy distribution characteristics were depicted not only from a time view but also from a frequency view. Abundant information, which cannot be made visible by FFT, such as transient frequency, transient energy, group delay, and so on, is distilled by the time–frequency image of the PWM drive pulses.

References

[1] Folland, G.B. and Sitaram, A. (1997) The uncertainty principle: a mathematical survey. *J. Fourier Anal. Appl.*, **3**(3), 207–238.

[2] Lobos,T., Rezmer, J., and Schegner, P. (2003) Parameter estimation of distorted signal using Prony method. IEEE Power Tech Conference Proceedings, Bologna, Italy, pp. 23–26.

[3] Marple, S.L., Jr., (1989) A tutorial overview of modern spectral estimation. International Conference on, Acoustics, Speech, and Signal Processing, Glasgow, UK, Vol. **4**, pp. 2152–2157

[4] Liaw, C.M., Lin, Y.M., Wu, C.H. and Hwu, K.I. (2000) Analysis, design, and implementation of a random frequency PWM inverter. *IEEE Trans. Power Electron.*, **15**(5), 843–854.

[5] Trzynadlowski, A.M., Blaabjerg, F., Pedersen, J.K. *et al.* (1994) Random pulse width modulation techniques for converter fed drive systems: a review. *IEEE Trans. Ind. Appl.*, **30**(5), 1166–1175.

[6] Lobos, T., Leonowicz, Z., and Rezmer, J. (2000) Harmonics and interharmonics estimation using advanced signal processing methods. Proceeding of the 9th IEEE International Conference on Harmonics Quality Power, Orlando, FL, pp. 335–340.
[7] Knezevic, J., Katic, V., and Graovac, D. (2000) Transient analysis of AC/DC converters input waveforms using wavelet. 10th Mediterranean Conference on Electrotechnical Conference, Vol. **3**, pp. 29–31.
[8] Tse, K.K., Chung, H.S.-H., Huo, S.Y. and So, H.C. (2000) Analysis and spectral characteristics of a spread-spectrum technique for conducted EMI suppression. *IEEE Trans. Power Electron.*, **15**(2), 399–410.
[9] Kuisma, M. (2003) Variable frequency switching in power supplies EMI-control: an overview. *IEEE Aerosp. Electron. Syst. Mag.*, **18**(12), 18–21.
[10] Santolaria, A., Balcells, J., Gonzalez, D., and Gago, J. (2003) Evaluation of switching frequency modulation in EMI emissions reduction applied to power converters. 29th Annual of IEEE (IECON), Industrial Electronic Socity, Vol. **3**, pp. 2306–2311.
[11] Vilathgamuwa, M., Deng, J. and Tseng, K.J. (1999) EMI suppression with switching frequency modulated DC–DC converters. *IEEE Ind. Appl. Mag.*, **5**(6), 27–33.
[12] Paramesh, J. and Jouanne, A.V. (2001) Use of sigma–delta modulation to control EMI from switch-mode power supplies. *IEEE Trans. Ind. Electron.*, **48**(1), 111–117.
[13] Tanaka, T., Ninomiya, T., and Harada, K. (1989) Random-switching control in DC-to-DC converters. 20th Annual IEEE, Power Electronics Specialists Conference, Milwaukee, WI, pp. 500–507.
[14] Zhang, Q., Wang, H.L., Zhu, Z. *et al.* (2004) *Theory and Application of Bifurcation and Chaos*, Tianjing University Press, China (in Chinese).
[15] Deng, Y. and Liang, Z. (1992) *Random Point Process and its Application*, Science Press, China (in Chinese).
[16] Crupper, J. and Schwarz, W. (2005) A performance estimation method for chaotic spread spectrum clock process. *IEEE Trans. Circuits Syst.*, **52**(4), 3383–3386.
[17] Orabi, M. and Ninomiya, T. (2003) Nonlinear dynamics of power-factor correction converter. *IEEE Trans. Ind. Electron.*, **50**(6), 1116–1125.
[18] Tse, K.K., Chung, H.S.-H., Huo, S.Y. and So, H.C. (2000) Analysis and spectral characteristics of a spread-spectrum technique for conducted EMI suppression. *IEEE Trans. Power Electron.*, **15**(2), 399–410.
[19] Sathiakumar, S., Shrivastava, Y., and Yim-Shu, L. (2004) Randomized switching schemes for DC-DC power conversion with reduced average switching frequency. 2004 International Conference on Power System Technology, 2004. PowerCon 2004, pp. 193–198.
[20] Sangswang, A. and Nwankpa, C.O. (2003) Random noise in switching DC-DC converter: verification and analysis. Proceedings of the 2003 International Symposium on Circuits and Systems, Vol. **3**, pp. 427–430.
[21] Tse, K.K., Chung, H.S.-H., Hui, S.Y.R., and So, H.C. (1999) Spectral characteristics of random carrier frequency switching in off-line switched mode power supply. Proceeding of the IEEE Appllied Power Electronics Conference Exposed, Dallas, TX, pp. 139–145.
[22] Tse, K.K., Ng, R.W.-M., Chung, H.S.-H. and Hui, S.Y.R. (2003) An evaluation of the spectral characteristics of switching converters with chaotic carrier frequency modulation. *IEEE Trans. Ind. Electron.*, **50**(1), 171–182.
[23] Gabor, D. (1946) Theory of communication. *J. Inst. Electr. Eng. - Part III: Radio Commun.*, **93**(26), 429–441.
[24] Cohen, R. (2012) Signal Denoising Using Wavelets.
[25] Rioul, O. and Vetterli, M. (1991) Wavelets and signal processing. *IEEE Signal Process Mag.*, **8**(4), 14–38.
[26] Gao, R.X. and Yan, R. (2011) *Wavelets: Theory and Applications for Manufacturing*, Springer Science, Boston, MA.
[27] Mallat, S. (2008) *DWT A Wavelet Tour of Signal Processing*, 3rd edn, Academic Press.

[28] Mallat, S. (1989) A theory for multiresolution signal decomposition: the wavelet representation. *IEEE Trans. Pattern Anal. Mach. Intell.*, **11**(7), 674–693.
[29] Meyer, Y. (1989) *Wavelets*, Springer-Verlag, Berlin.
[30] Daubechies, I. (1992) *Ten Lectures on Wavelets*, SIAM, Philadelphia, PA.
[31] Hubbard, B.B. (1998) *The World According to Wavelets*, A K Peters, Ltd.
[32] Rioul, O. and Vetterli, M. (1991) Wavelets and signal processing. *IEEE Signal Process. Mag.*, **8**(4), 14–38.
[33] Weiss, L.G. (1994) Wavelets and wideband correlation processing. *IEEE Signal Process. Mag.*, **11**(1), 13–32.
[34] Mallat, S. (2002) *A Wavelet Tour of Signal Processing*, Academic Press, San Diego, CA.
[35] Jawerth, B. and Sweldens, W. (1994) An overview of wavelet based multiresolution analyses. *SIAM Rev.*, **36**(33), 377–412.
[36] Yang, R., Zhang, B. and Qiu, D.Y. (2009) Time–frequency and wavelet transforms of EMI dynamic spectrum in chaotic converter. *IEEE Trans. Power Electron.*, **24**(4), 1083–1092.

9

Dynamic Chaos Spectrum of Chaotic Switching Converters II: Prony Method

9.1 Introduction

Recently, research has shown that the chaotification method can be used to reduce the electromagnetic interference (EMI) of DC–DC converters. It is well known that the EMI of electronic equipment is estimated by its spectrum. Conventionally, the fast Fourier transform (FFT) is used to analyze the spectrum. However, it is not applicable to the inter-harmonics, the noninteger harmonics of the main fundamental frequency under consideration. But these are prominent features of chaotic signals.

Traditionally, the strength of the EMI is measured by an estimation of the system harmonics, namely, by calculating the power spectral density (PSD) based on FFT. This spectral analysis approach is computationally efficient and can provide reasonable results for signal processes in most cases. However, PSD estimation based on FFT has some drawbacks that we mentioned last chapter. The most prominent one is that of frequency resolution, that is, the ability to distinguish the spectral responses of two or more signals. The frequency resolution in hertz is roughly the reciprocal of the time interval in seconds, over which sampled data is available. The second one is due to the implicit windowing of the data that occurs when processing with the FFT. Windowing manifests itself as "leakage" in the spectral domain, that is, energy in the main lobe of a spectral response "leaks" into the side-lobes, obscuring and distorting other nearby spectral responses that are presented in [1]. These two drawbacks limit the application of FFT in analyzing short data records, because many measured processes are brief or have slowly time-varying spectra that are often considered as constant in short sampling intervals. Further, FFT cannot efficiently estimate inner-harmonics, since it assumes the harmonics to be integral multiples of the fundamental frequency [2].

Chaos Analysis and Chaotic EMI Suppression of DC-DC Converters, First Edition. Bo Zhang and Xuemei Wang.

To alleviate the limitations of the FFT, several new modern spectral estimation methods have been proposed [3–5]. Prony analysis is an emerging methodology, which extends Fourier analysis by directly estimating the frequency, damping, strength, and relative phase of the modal components presented in a given signal [6].

In this chapter, the Prony method is employed for the chaotic spectral estimation of DC–DC converters, and numerical simulations show its advantages over the traditional FFT.

9.2 Prony Method

Traditional spectrum analysis methods such as Fourier transform are commonly used to estimate the stationary signal but are no use for time-varying signals. Prony analysis was developed by Gaspard Riche and Baron de Prony in 1795 in order to explain the expansion of various gases [7–9]. The basic idea of the Prony method is to model a linear combination of damped complex exponentials to the signal that are uniformly sampled [10, 11]. Although this method is not a spectrum estimation technique, it has a close relationship to the autoregressive moving average (ARMA) parameter estimation and the least-squares linear prediction algorithms used for autoregressive (AR). The Prony method tries to fit a deterministic exponential model to the data, in contrast to AR and ARMA methods that seek to fit a random model to the second-order data statistics [10].

According to this method, the modal parameters of the respective frequency components of the amplitude, phase, frequency, and attenuation factor may be analyzed, which is also the advantage of Prony analysis compared with other oscillatory signal analysis techniques. In addition, it best fits a reduced-order model to a high-order system both in time and frequency domains [9]. Table 9.1 lists the major differences between Fourier series and Prony analysis. The Prony method improves the frequency resolution and is not affected by windowing. Thus, the Prony method can be applied not only to the spectral estimation but also to obtaining information about amplitudes, phases, frequencies, and damping factors of harmonics. Furthermore, it is shown that the Prony method can be used to reconstruct or to fit the sampled data.

Table 9.1 Comparison between Fourier series and Prony analysis [12]

Fourier series (FS)	Prony analysis (PA)
Steady-state signal, time-invariant	Dynamic signal, time varying
FS fits a sum of un-damped complex exponentials	PA fits a sum of damped complex exponentials
FS computes amplitude, phase, and frequency of the signal components	Apart from amplitude, phase, and frequency, PA also computes damping coefficients of the signal components

9.2.1 Basic Principle of Prony Method

Prony analysis is a method of fitting a linear combination of exponential terms to a signal as shown in Equation 9.1 [10].

$$x(t) = \sum_{i=1}^{P} A_i e^{\alpha_i t} \cos(2\pi f_i t + \theta_i) \tag{9.1}$$

Where P is the total number of damped exponential components, A_i is the amplitude of components i, α_i is the damping factor of components i, θ_i is the phase of components i, and f_i is the frequency of components i.

Using Euler's theorem and letting $t=nT$, the samples of $x(t)$ at $t=nT$ may be rewritten as

$$\hat{x}(n) = \sum_{i=1}^{P} b_i z_i^n \tag{9.2}$$

where

$$b_i = \frac{1}{2} A_k e^{j\theta_i}$$

$$z_i = e^{(\alpha_i + j2\pi f_i)T}$$

and T is sampling period, $\hat{x}(n)$ is the fitted value of measured data of $x(t)$ and consists of N samples.

Constructing a cost function, we obtain

$$\varepsilon = \sum_{n=0}^{N-1} \left| x(n) - \hat{x}(n) \right|^2 \tag{9.3}$$

If the sum of the squared errors between the measured data $x(n)$ and the fitted value $\hat{x}(n)$ is minimum, the parameters $\{A_i, \alpha_i, f_i, \theta_i\}$ may be derived. But this is a complex nonlinear solving problem, due to the existence of the exponential items in Equation 9.2. It can be settled down by using the Prony algorithm whose characteristic is using a linear differential equation instead of solving the nonlinear equations. In order to derive the linear differential equation, Prony proposed to define the characteristic polynomial whose roots are exponential terms z_i

$$F(z) = \prod_{i=1}^{P} (z - z_i) = \sum_{i=0}^{P} a_i z^{P-i} \tag{9.4}$$

where a_i is the coefficient and $a_0 = 1$. Prony proved that the solutions are homogeneous between the constant coefficient linear differential equation and Equation 9.1.

Shifting n in Equation 9.2 from n to $n - k$, we have

$$\hat{x}(n-k) = \sum_{i=1}^{P} b_i z_i^{(n-k)} \qquad 0 \le n-k \le N-1 \tag{9.5}$$

Multiplying the parameter a_i, then

$$\sum_{k=0}^{P} a_k\hat{x}(n-k) = \sum_{i=1}^{P} b_i \sum_{k=0}^{P} a_k z_i^{(n-k)} \quad p \le n \le N-1 \tag{9.6}$$

For $z_i^{n-k} = z_i^{n-P} z_i^{P-k}$, Equation 9.6 can be modified to

$$\sum_{k=0}^{P} a_k\hat{x}(n-k) = \sum_{i=1}^{P} b_i z_i^{(n-P)} \sum_{k=0}^{P} a_k z_i^{(P-k)} \tag{9.7}$$

Note that the right second summation of Equation 9.7 may be recognized as the polynomial defined by Equation 9.3, and evaluated at each of its roots z_i yielding the zero result

$$\sum_{k=0}^{P} a_k\hat{x}(n-k) = 0 \tag{9.8}$$

Then, Equation 9.2 yields the backward recursive difference equations

$$\hat{x}(n) = -\sum_{i=1}^{P} a_i\hat{x}(n-i) \quad P \le n \le N-1 \tag{9.9}$$

Equation 9.9 can be considered as a set of linear equations whose unknown coefficients are $\{A_i, \alpha_i, f_i, \theta_i\}$. In other words, $\{A_i, \alpha_i, f_i, \theta_i\}$ may be obtained by solving Equation 9.9. Therefore, the key problem of Prony analysis is that Equation 9.2 is the homogeneous solution of linear difference equations and the coefficients a_i in Equation 9.9 are related to the characteristic polynomial Equation 9.4.

Supposing that the errors between sampled data and approximations are

$$e(n) = x(n) - \hat{x}(n) \quad 0 \le n \le N-1 \tag{9.10}$$

So Equation 9.9 can be written as

$$\begin{aligned} x(n) &= -\sum_{i=1}^{P} a_k \cdot x(n-i) + \sum_{i=0}^{P} a_i \cdot e(n-i) \\ &= -\sum_{i=1}^{P} a_i \cdot x(n-i) + u(n) \end{aligned} \tag{9.11}$$

where $u(n) = \sum_{i=0}^{P} a_k \cdot e(n-i)$ and $a_0 = 1,\ n = P, \cdots, N-1$. Here, $x(n)$ is regarded as the output of the Pth-order AR model driven by noise $u(n)$, and the variables a_i are the coefficients of AR model. Solving the canonical equation of AR model may yield the parameters a_i $(i = 1, 2, \ldots, P)$.

If the errors are approximately Gaussian, we will try to estimate or identify the unknown parameters in the exponential or sinusoid signal by minimizing the total squared errors of Equation 9.3 as

$$\sum_{n=P}^{N-1} |e(n)|^2 \tag{9.12}$$

This becomes a nonlinear least squares problem of the unknown parameters *bi* and *zi*. If we define a cost function

$$J(a) = \sum_{n=P}^{N-1} |u(n)|^2 \tag{9.13}$$

then a set of linear matrix equations may be obtained as

$$\begin{bmatrix} x(p) & x(p-1) & \dots & x(0) \\ x(p+1) & x(p) & \dots & x(1) \\ \vdots & \vdots & & \vdots \\ x(N-1) & x(N-2) & \cdots & x(N-p-1) \end{bmatrix} \begin{bmatrix} 1 \\ a_1 \\ \vdots \\ a_p \end{bmatrix} = \begin{bmatrix} u(p) \\ u(p+1) \\ \vdots \\ u(N-1) \end{bmatrix} \tag{9.14}$$

In order to minimize the cost function, Equation 9.13 can be written as

$$J(a) = \sum_{n=P}^{N-1} |u(n)|^2 = \sum_{n=p}^{N-1} \left| \sum_{j=0}^{p} a_j e(n-j) \right|^2 \tag{9.15}$$

Let $\dfrac{\partial J}{\partial a_j} = 0$, then we get

$$\sum_{j=0}^{p} a_j \left[\sum_{n=p}^{N-1} x(n-j)\, x^*(n-i) \right] = 0 \qquad i = 1, 2, \cdots, p \tag{9.16}$$

The corresponding minimum error energy is

$$u_p = \sum_{j=0}^{p} a_j \left[\sum_{n=p}^{N-1} x(n-j)\, x^*(n) \right] \tag{9.17}$$

Defining

$$r(i,j) = \sum_{n=p}^{N-1} [x(n-j)x^*(n-i)] \tag{9.18}$$

So Equations 9.16 and 9.17 may be combined into the normal equation:

$$\begin{bmatrix} r(0,0) & r(0,1) & \dots & r(0,p) \\ r(1,0) & r(1,1) & \dots & r(1,p) \\ \vdots & \vdots & & \vdots \\ r(p,0) & r(p,1) & \cdots & r(p,p) \end{bmatrix} \begin{bmatrix} 1 \\ a_1 \\ \vdots \\ a_p \end{bmatrix} = \begin{bmatrix} u(p) \\ 0 \\ \vdots \\ 0 \end{bmatrix} \tag{9.19}$$

Solving Equation 9.19 yields the parameters a_i ($i = 1, 2, \dots, P$). Once a_i are computed, then re-substituting a_i into the characteristic polynomial Equation 8.4, and z_i into the roots of Equation 8.4, there may be determined

$$\begin{cases} f_i = \arctan\left[I_m/R_e\left(z_k\right)\right]/2\pi\Delta t & i = 1, 2, \cdots, P \\ \alpha_i = \ln|z_i|/\Delta t \end{cases} \tag{9.20}$$

Wherein the order number P can be determined using the singular value decomposition (SVD) [13, 14] method of AR models.

According to Equation 9.5, in order to obtain b_i, let $n = 0, 1, \dots, N-1$, and we resolve the following matrix form

$$\mathbf{Zb} = \mathbf{x} \tag{9.21}$$

where $\mathbf{Z}$ is a Van der Monde matrix, and $\mathbf{Z} = \begin{pmatrix} 1 & 1 & \dots & 1 \\ z_1 & z_2 & \dots & z_P \\ \vdots & & \ddots & \vdots \\ z_1^{N-1} & z_2^{N-1} & \dots & z_P^{N-1} \end{pmatrix}$, $\mathbf{b} = \begin{bmatrix} b_1 \\ b_2 \\ \vdots \\ b_P \end{bmatrix}$,

$\mathbf{x} = \begin{bmatrix} x(0) \\ x(1) \\ \vdots \\ x(N-1) \end{bmatrix}$.

Using the least squares method on Equation 9.21, a well-known solution is:

$$\mathbf{b} = (\mathbf{Z}^H\mathbf{Z})^{-1}\mathbf{Z}^H\mathbf{x} \tag{9.22}$$

in which $\mathbf{Z}^{\mathrm{H}}$ stands for the conjugate transpose matrix of $\mathbf{Z}$.

Finally, according to Equation 9.2, the amplitudes A_{i} and the phase θ_{i} may be obtained as

$$\begin{cases} A_i = |b_i| & i = 1, 2, \cdots, P \\ \theta_i = \arctan\left[I_m(b_i)/R_e(b_i)\right] \end{cases} \tag{9.23}$$

Thus, $\hat{x}(n)$ ($n = 0, 1, \dots, N-1$) are obtained and denoted in vector form as $\hat{x}$.

Denote the Fourier transform of $\hat{x}$ as $\hat{X}(f)$. Then the PSD of the N sampled data $\hat{P}_{\mathrm{prony}}(f)$ can be expressed as

$$\hat{P}_{Prony}(f) = |\hat{X}(f)|^2 \tag{9.24}$$

where

$$\hat{X}(f) = \sum_{k=1}^{P} A_k e^{j\theta} \frac{2\alpha_k}{|\alpha_k|^2 + (2\pi(f - f_k))^2} \tag{9.25}$$

9.2.2 *Classical Computing Process of Prony Analysis*

There are the three steps in classical Prony analysis [12, 15].

9.2.2.1 Solve the Linear Prediction Model, Which is Constructed by the Observed Data Set

First, the coefficients of a linear prediction model (LPM) are calculated. LPM assumes that $x(n)$ is a linear combination of the past P outputs (that is, it can be "linearly predicted" from these outputs) for each output sample of a signal. Then, the coefficients are constant from sample to sample

$$x(n) = a_1 x(n-1) + a_2 x(n-2) + \cdots + a_P x(n-P) \tag{9.26}$$

Estimation of LPM coefficients a_n is crucial for the derivation of the frequency, damping, magnitude, and phase angle of a signal. To estimate these coefficients accurately, many algorithms can be used. A matrix representation of the signal at various sample times can be formed by sequentially writing the linear prediction of $x(n)$ repetitively. By inverting the matrix representation, the linear coefficients a_n can be derived from

$$\begin{bmatrix} x_P \\ x_{P+1} \\ \vdots \\ x_{n-1} \end{bmatrix} = \begin{bmatrix} x_{P-1} x_{P-2} & \cdots & x_0 \\ x_{P+1} x_{P-1} & \cdots & x_1 \\ \vdots \; \vdots & & \\ x_{n-2} x_{n-3} & \cdots & x_{n-P-1} \end{bmatrix} \begin{bmatrix} a_1 \\ a_2 \\ \vdots \\ a_P \end{bmatrix} \tag{9.27}$$

An algorithm, which uses SVD for the matrix inversion to derive the LPM coefficients, is called an SVD algorithm [12].

9.2.2.2 Find the Roots of the Characteristic Polynomial Formed from the Linear Prediction Coefficients

The characteristic polynomial of the LPM Equation 9.27 is as Equation 9.4 motioned above and can be represented as

$$z^P - a_1 z^{P-1} - \cdots - a_{P-1} z - a_P = (z - \hat{z}_1)(z - \hat{z}_2) \cdots (z - \hat{z}_P) \tag{9.28}$$

As coefficients a_n are known from Equation 9.27, the roots z_n of polynomial Equation 9.28 can be rapidly computed.

9.2.2.3 Solve the Original Set of Linear Equations to Yield the Estimates of the Exponential Amplitude and Sinusoidal Phase

In this step, the magnitudes and the phase angles of the signal are solved in the least squares sense. According to Equations 9.21–9.23, the magnitude A_i and phase angle θ_i are thus calculated.

9.3 Estimating PSD Using the Prony Method

It is known that the frequency resolution of FFT is proportional to $1/N\Delta t$, where Δt is the sampled period. DC–DC converters always work at a relatively high frequency, thus the resolution of FFT is not satisfactory in practice [16]. In addition, for the case of short data obtained in failure diagnosis, the resolution of FFT is also very low because N is small.

However, the Prony method overcomes these drawbacks, except its computation is a little more complex than that of FFT. It can be used to estimate the PSD of DC–DC converters, especially when converters work in chaotic mode.

To illustrate the effectiveness of the Prony method in improving the frequency resolution of the PSD compared with the FFT method, a periodic signal is taken as an example to estimate its PSD

$$y(t) = \sin(2\pi f_1 t + 0.5\pi) + 0.6\cos(2\pi f_2 t + 0.2\pi) + 2\sin(2\pi f_3 t) \tag{9.29}$$

where $f_1 = 200$ Hz, $f_2 = 198$ Hz, and $f_3 = 75$ Hz. Let $N = 256$ be the number of sampled data, and $f_S = 2$ kHz be the sampling frequency.

Figure 9.1 shows the PSD estimation, and the related coefficients at $P = 8$ are given in Table 9.2.

Investigating A_i and f_i in Table 9.2, it can be seen that by discarding the negative frequencies and values of A_i that are correspondingly small, only three positive frequencies, that is, $f = 75$ Hz, $f = 198$ Hz, and $f = 200$ Hz, remain. This solution is consistent with Equation 9.28. Further investigating the damping factors α_i, it is noted that the three damping factors corresponding to the three positive frequencies are very small. This implies that the corresponding signals in the polynomial exponential function Equation 9.1 are periodic, while the others with big damping factors are instant.

As a comparison, the N-point FFT is adopted, where $N = 256$ and the frequency resolution is $\Delta f = fs/N = 7.8125$ Hz. This means, if $|f_1 - f_2| \le \Delta f$, the FFT is not able

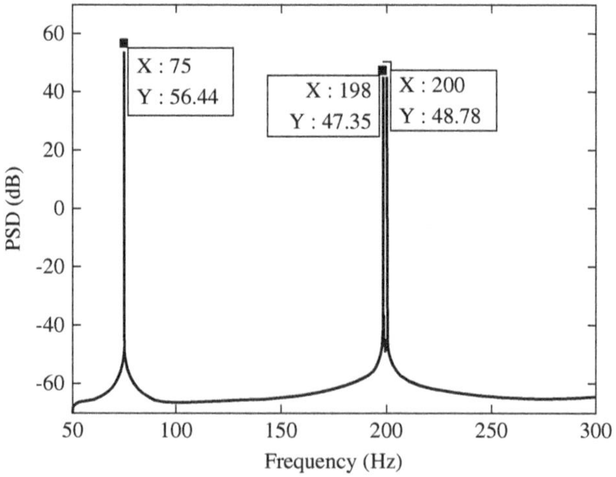

Figure 9.1 PSD estimation via Prony method

Table 9.2 Coefficients derived at $P = 8$

f_i	A_i	α_i	θ_i
75	2	2.22e-12	−1.81e-15
198	0.6	−6.26e-11	0.6283
200	1	5.17e-11	−0.9425
375.12	1.14e-12	−214.31	1.9514
499.55	2.16e-12	−372.43	2.3287
618.04	1.63e-12	−325.45	2.7256
756.36	2.17e-12	−395.29	2.5592
873.27	2.64e-12	−442.43	3.0456

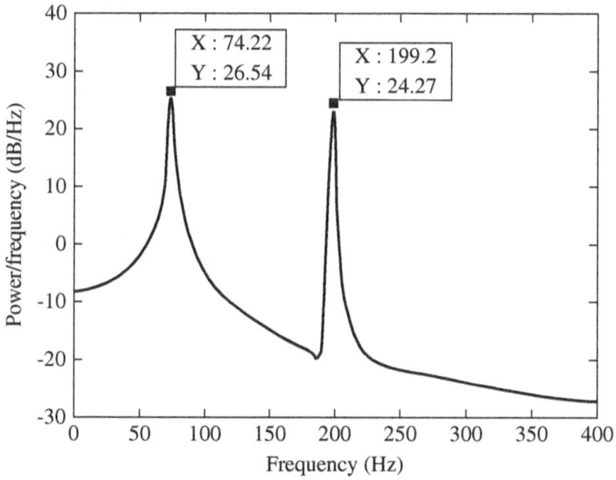

Figure 9.2 PSD estimate via FFT

to distinguish these two frequencies. It is shown in Figure 9.2 that only two peaks, $f_1 = 74.22$ Hz and $f_2 = 199.2$ Hz, are identified using the FFT.

It is remarked that although the FFT method is simple and computationally effective, its frequency resolution is low, especially for short sampled data sequences. In contrast, the Prony method has merits in improving frequency resolution and data reconstruction. In particular, due to the existence of rich inner-harmonics and random-like behavior in chaotic systems, the Prony method is more powerful and effective than the FFT method.

In addition, the Prony method can be employed to reconstruct the sampling data using the obtained parameters, A_i, α_i, f_i, and θ_i, as shown in Figure 9.3. It is obvious that the errors between the original signal and the reconstructed one are very small (Figure 9.4) and the errors only reach to the order of 10^{-11}.

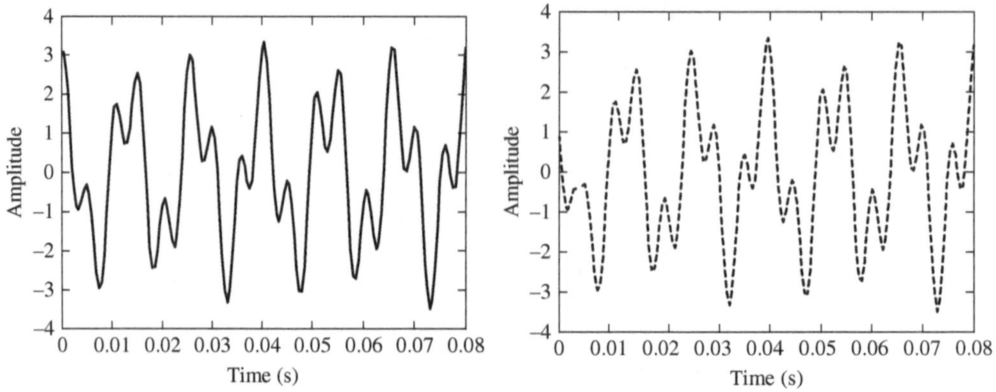

Figure 9.3 Original signal (a) and the reconstructed signal (b)

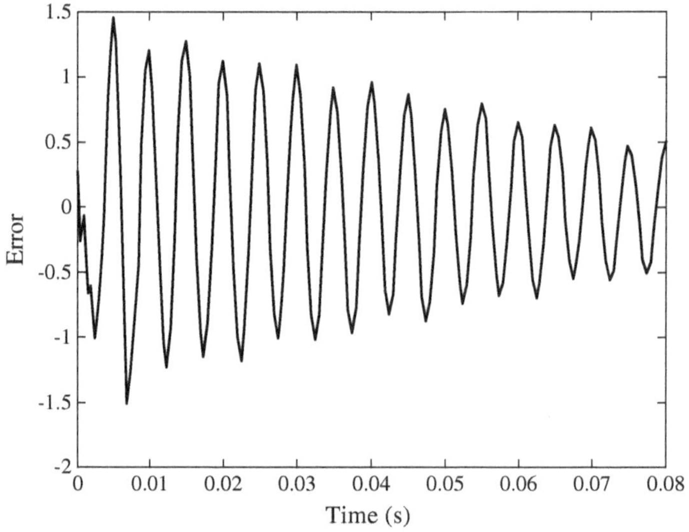

Figure 9.4 Errors between the original signal and the reconstructed signal

9.4 Chaotic Spectral Estimation of DC–DC Converters Based on the Prony Method

It is known that DC–DC converters produce EMI and thus result in electromagnetic pollution [16]. With the increase in the usage of electronic equipment, the EMI problem has attracted increasing attention from engineers [17]. Recently, studies have shown that DC–DC converters have a broadband spectrum when they operate in chaotic modes, and the energy of EMI is more uniformly distributed in the frequency band [18]. Thus, the peak values of EMI can be decreased, but rich inner-harmonics are generated. The inner-harmonics may result in quality degradation of the transmission energy, increase of power loss, reliability degradation of the converter systems, and so on [19, 20]. Thus, it is meaningful to investigate the inner-harmonics by control systems.

The traditional FFT can only estimate the fundamental frequency and its integral multiples, so it is not applicable for inner-harmonics. Instead, the Prony method is employed here for the spectral estimation of the inductor current of the boost converter, whose circuitry is shown in Figure 1.9. Therein, the reference current I_{ref} serves as the control parameter. By adjusting the reference current, the boost converter may exhibit period-doubling bifurcation, tangent bifurcation, and chaotic oscillations. In the following, the Prony method is used for spectral estimation of the inductor current corresponding to the three operating modes [16].

Assume the circuit parameters are $E = 10\,\text{V}$, $L = 1\,\text{mH}$, $C = 12\,\mu\text{F}$, $R = 20$, and $f_c = 10\,\text{kHz}$, where E is the input voltage, L the input inductor, C the output capacity, R load resistance, and f_c the clock frequency. This converter operates in CCM (continuous conduction mode).

In simulation waveforms of the inductor current, we can find that the converter exhibits period-1 (Figure 9.5a), period-2 (Figure 9.5b), chaotic behaviors (Figure 9.5c), and period-3 (Figure 9.5d) at $I_{ref} = 1, 1.8, 4, 4.55$ A respectively.

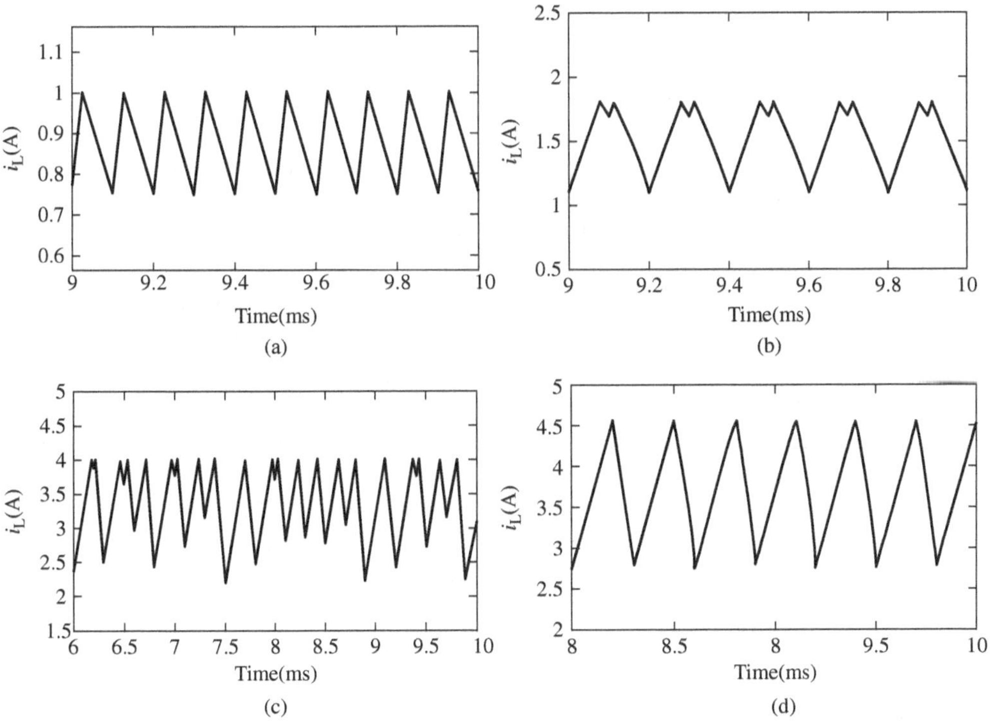

Figure 9.5 Sampling current waveform at (a) $I_{ref} = 1$ A, (b) $I_{ref} = 1.8$ A, (c) $I_{ref} = 4$ A, and (d) $I_{ref} = 4.55$ A

Table 9.3 Coefficients derived at $P = 20$ of period-2

f_i	A_i	α_i	θ_i
2537	1.234e-11	−2756	2.8402
5000	0.3023	0	3.1394
7660	3.634e-12	−3385	−1.8412
10 000	0.0541	0	2.9843
12 543	2.087e-12	−2447	1.9705
15 000	0.0051	0	−2.9813
17 184	2.380e-11	−89 710	3.034
17 278	7.911e-13	−2170	−1.5849
20 000	0.0323	0	−3.0452
21 985	8.103e-13	−4277	2.137
25 000	0.0083	0	0.7081
29 036	1.068e-12	−8259	1.1416
30 000	0.0197	0	−2.8468
32 832	8.994e-13	−2898	−0.3325
35 000	0.0056	0	1.1747
37 522	6.213e-13	−2141	2.8715
40 000	0.0109	0	−2.7971
42 189	6.422e-13	−2446	−0.8548
45 000	0.0024	0	2.2263
46 593	1.072e-12	−6141	−2.4718

In order to estimate the spectrum by the Prony method introduced above, we assume that $I_{ref} = 1.8$ A corresponding to the period-2 mode and $P = 20$, which is an empirical value. Using the Prony method, the coefficients can be derived in Table 9.3 after omitting the negative frequencies.

We can find from Table 9.3 that the direct current (DC) component 0.0541 A with $f_i = 1000$ Hz and the alternating current (AC) components with non-zero values of f_i can be decomposed. That is, by investigating f_i, one can distinguish not only the fundamental frequency and its integral multiple but also the inner-harmonics. By observing A_i, the amplitude of the fundamental frequency component is the largest one among all the AC components.

The simulation results of the spectral estimation using the Prony method are illustrated in Figures 9.6a, 9.7a, 9.8a, and 9.9a with $N = 512$, $P = 80$. For comparison, a similar simulation using the FFT method is also carried out and the results are shown in Figures 9.6b, 9.7b, 9.8b, and 9.9b. It is obvious that the Prony method can much more accurately locate the frequencies of the harmonics corresponding to the peaks for all cases.

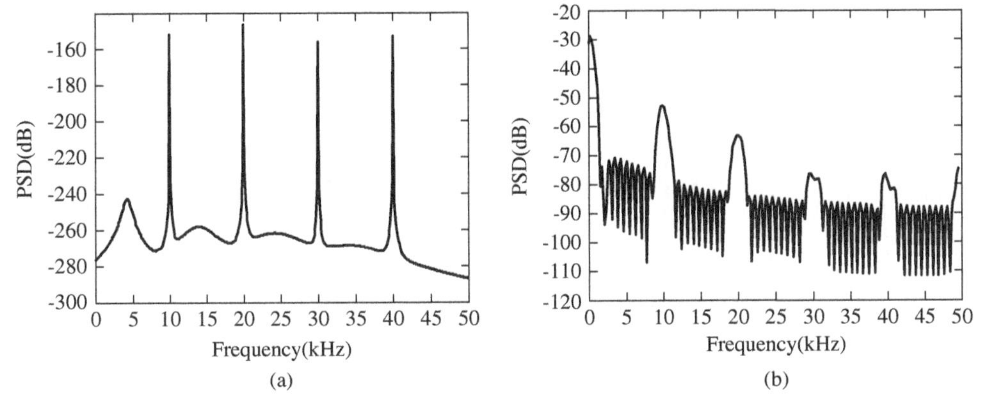

Figure 9.6 PSD estimation by (a) Prony and (b) FFT with $I_{ref} = 1$ A

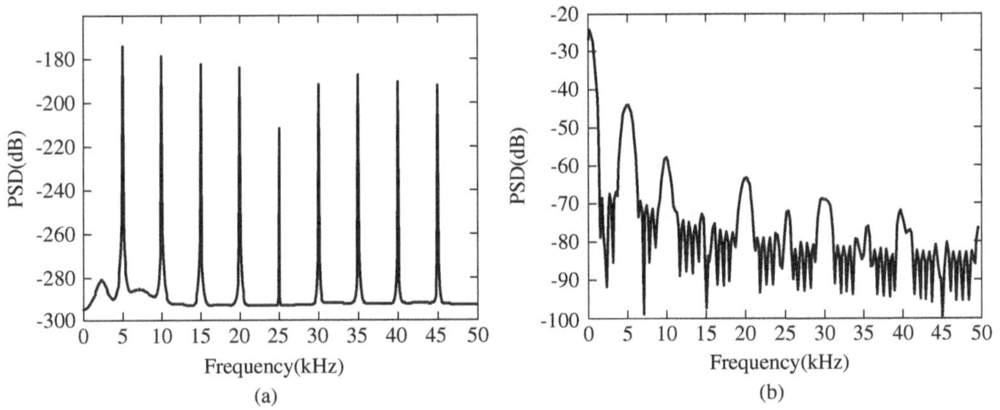

Figure 9.7 PSD estimation by (a) Prony and (b) FFT with $I_{ref} = 1.8$ A

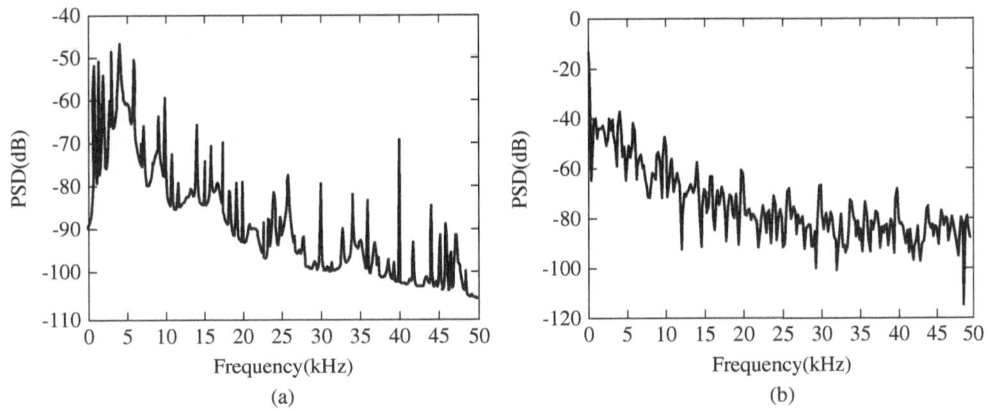

Figure 9.8 PSD estimation by (a) Prony and (b) FFT with $I_{ref} = 4$ A

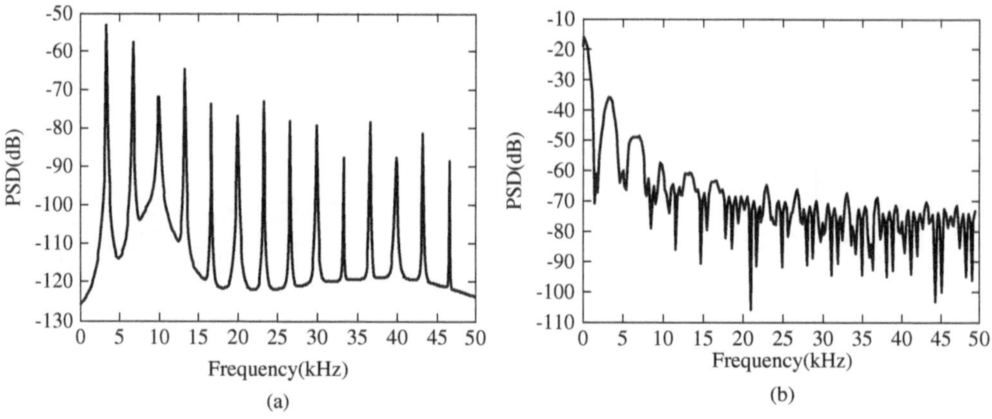

Figure 9.9 PSD estimation by (a) Prony and (b) FFT with $I_{\text{ref}} = 4.55$ A

9.5 Conclusions

Traditionally, the strength of the EMI has been measured by the estimation of the system harmonics; namely, by calculating the PSD based on FFT. In this chapter, the Prony method has been employed in the chaotic spectral estimation of DC–DC converters, and numerical simulations show its advantages over the traditional FFT.

References

[1] Marple, S.L. (1989) A tutorial overview of modern spectral estimation. Processing, International Conference on Acoustics, Speech, and Signal, Glasgow, UK, Vol. **4**, pp. 2152–2157.

[2] Leonowicz, Z., Lobos, T., and Rezmer, J. (2000) Harmonics and interharmonics estimation using advanced signal processing methods. Proceeding of the 9th IEEE International Conference on Harmonics and Quality of Power, Orlando, FL, pp. 335–340.

[3] McGranaghan, M.F., Zavadil, R.M., Hensley, G. *et al.* (1991) Impact of utility switched capacitors on customer systems-magnification at low voltage capacitors. Proceeding of the IEEE Power Engineering Society Conference Transmission and Distribution, Dallas, TX, pp. 908–914.

[4] Saiz, V.M.M. and Guadalupe, J.B. (1997) Application of Kalman filtering for continuous real-time tracking of power system harmonics. *IEE Proc. Gener. Transm. Distrib.*, **144**(1), 13–20.

[5] Grondin, R., Kamwa, I. and McNabb, D. (1996) On-line tracking of changing harmonics in stressed power systems: application to Hydro-Quebec network. *IEEE Trans. Power Delivery*, **11**(4), 2020–2027.

[6] Hauer, J.F. *et al.* (1990) Initial results in prony analysis of power system response signals. *IEEE Trans. Power Syst.*, **5**(1), 80–89.

[7] Prony, G.R.B. (1795) Essai éxperimental et analytique: ... (in French). *J. L'école Polytech. (Paris)*, **1**, 24–76.

[8] Hilderbrand, F.B. (1974) *Introduction to Numerical Analysis*, McGraw-Hill, New York.

[9] Marple, L. (1987) *Digital Spectral Analysis With Application*, Prentice-Hall, Englewood Cliffs, NJ.

[10] Leonowicz, Z., Lobos, T. and Rezmer, J. (2003) Advanced spectrum estimation methods for signal analysis in power electronics. *IEEE Trans. Ind. Electron.*, **50**(3), 514–519.

[11] Meunier, M. and Brouaye, F. (1998) Fourier transform, wavelets, prony analysis: tools for harmonics and quality of power. Proceeding of the 8th International Conference on Harmonics and Quality of Power, Athens, Greece, pp. 71–76.
[12] Singh, S. (2003) Application of Prony analysis to characterize pulsed corona reactor measurements. Master thesis, University of Wyoming.
[13] Golub, G.H. and Van Loan, C.F. (1996) *Matrix Computations*, The Johns Hopkins University Press.
[14] Ranade, A.G. (2007) Some Uses of Spectral Methods. http://www.cse.iitb.ac.in/~ranade (accessed 7 June 2014).
[15] Qi, L., Qian, L., Woodruff, S. and Cartes, D. (2007) Prony analysis for power system transient harmonics. *EURASIP J. Appl. Signal Process.*, **2007**(1), 170–181.
[16] Li, H., Li, Z., Halang, W.A. *et al.* (2007) Analyzing chaotic spectra of DC–DC converters using the prony method. *IEEE Trans. Circuits Syst. II*, **54**(1), 61–65.
[17] Bryant, B. and Kazimierczuk, M.K. (2006) Voltage loop of boost pwm dc-dc converters with peak current-mode control. *IEEE Trans. Circuits Syst. I*, **53**(1), 99–105.
[18] Deane, J.H.B. and Hamill, D.C. (1996) Improvement of power supply EMC by chaos. *Electron. Lett.*, **32**(12), 1045.
[19] Lee, J.H. and Kim, H.T. (2003) Selecting sampling interval of transient response for the improved prony method. *IEEE Trans. Antennas Propag.*, **51**(1), 74–77.
[20] Chang, C.L. and Bai, H. (1999) Experimental study on the performance of a single discharge wire-plate electrostatic precipitator with back corona. *J. Aerosol Sci.*, **30**(3), 325–340.

10

Chaotic PWM Suppressing EMI of Power Electronic Converters

10.1 Introduction

The peaks of electromagnetic interference (EMI) of power electronic converters mainly converge on frequencies which are the multiples of switching frequencies. These harmonics spread out through the transmission line and space electromagnetic field. Therefore, they lead to conducted interference and radiated interference problems. At present, the main measures for suppressing EMI of power electronic converters are as follows.

1. Add an input filter and shield technology.
2. Add an absorption circuit to the high-frequency semiconductor to reduce dv/dt and di/dt.
3. Improve the printed circuit board (PCB) design to reduce the area of high-frequency current loop.

All methods mentioned above may reduce EMI by adding circuit components, strongly depending on personal experience, so it is obvious that the EMI problem of power electronic converters cannot be solved radically.

Over the past 10 years, people have paid more attention to using pulse width modulation (PWM) technology to suppress the EMI of power electronic converters. The basic principle is using PWM to make EMI energy spread as equable as possible in the frequency domain, which can reduce the frequency spectrum peaks and realize a suppression of EMI. The main technologies for suppressing EMI are periodic frequency modulation, jitter frequency, random frequency modulation, and chaotic frequency modulation. Studies have found that compared with other modulation technologies, the chaotic modulated PWM signal can reduce harmonics efficiently and improve the distribution of the frequency spectrum of DC–DC converters. Kuisma [1] points out

Chaos Analysis and Chaotic EMI Suppression of DC-DC Converters, First Edition. Bo Zhang and Xuemei Wang.

that using a chaotic signal instead of a random signal does not reduce performance, but makes costs lower. Moreover, it has a better performance to some degree. Therefore, chaotic PWM has considerable development space both in theory and practice. However, presented studies of chaotic behavior are mostly expressed in abstract forms, so they are very difficult to apply to practical engineering design directly.

Therefore, through analyzing the spectrum characteristics of chaotic PWM suppressing EMI, this chapter provides a simple and easy-to-achieve chaos signal generation circuit-piecewise linear capacitor chaos circuit. Without changing the basic parameters and control circuit of primary power electronic converters, the signal from a chaos modulation circuit is directly sent to the clock port of PWM controlled ICs, such as UC3842 and SG3525. And then the chaotic PWM suppressing EMI is realized. This work will progress the engineering application of chaos.

10.2 The Principle of Chaotic PWM Suppressing EMI

10.2.1 Basic Theory of Frequency Modulation

Frequency modulation is that instantaneous frequency of high-frequency carrier changes with the varying of the low-frequency modulating signal [2, 3]. When the carrier instantaneously changes with the low-frequency sinusoidal signal, it is called sinusoidal modulation. Then the time-domain signal v_{FM} can be expressed by,

$$v_{\text{FM}}(t) = A_{\text{c}} \cos \left[2\pi f_{\text{c}} t + \frac{k_f A_{\text{m}}}{f_{\text{m}}} \sin 2\pi f_{\text{m}} t \right] \tag{10.1}$$

where A_{c} is the amplitude and f_{c} is frequency, and this carrier signal is a cosine signal. The instantaneous frequency of carrier signal may be modulated by frequency $f_{\text{m},}$ amplitude A_{m}, and proportionality coefficient k_f of the modulating signal.

To describe frequency modulating characteristics, the modulation index m_f and modulation rate are defined by

$$m_f = \frac{k_f A_{\text{m}}}{f_{\text{m}}} = \frac{\Delta f}{f_{\text{m}}} \tag{10.2}$$

$$\delta = \frac{\Delta f}{f_{\text{c}}} \tag{10.3}$$

where the Δf determined bandwidth of signal energy is the maximum frequency deviation of carrier frequency.

Based on Bessel function theory, after frequency modulating, the time-domain expression can be defined as

$$v_{\text{FM}}(t) = A_{\text{c}} \sum_{n=-\infty}^{\infty} J_n(m_f) \cos(\omega_{\text{c}} + n\omega_{\text{m}})t \tag{10.4}$$

where $J_n(m_f) = \frac{1}{2\pi}\int_0^{2\pi} e^{jm_f \sin \omega_m t} e^{-jn\omega t} d(\omega t)$ is first kind nth order Bessel function of modulation index m_f.

Making a Fourier transform of Equation 10.4, the frequency-domain of the modulation signal can be derived by

$$S_{FM}(\omega) = \pi A_c \sum_{n=-\infty}^{\infty} J_n(m_f)[\delta(\omega - \omega_c - n\omega_m) + \delta(\omega + \omega_c + n\omega_m)] \qquad (10.5)$$

Frequency modulation belongs to nonlinear modulation, and its spectrums are very complicated. The main features of frequency modulation are as follows.

1. Theoretically, there are infinite frequency components in a spectrum; this means that the modulation wave spectrum can be broadened.
2. When $n = 0$, the spectrum is just the carrier's frequency f_c, where its amplitude is $A_c J_0(m_f)$. When $n \neq 0$, the side frequency components $f_c + nf_m$ distribute on both sides of f_c. The interval between spectrum line is f_m, and amplitude is $A_c\ J_n(m_f)$.
3. The spectrum is the nonlinear relationship between the modulation wave and modulating signal.

Research has shown that the practical variable frequency range $J_n(m_f)$ will sharply decline with increasing n. The effective bandwidth of the modulation wave can be calculated by the Carson formula

$$B_{FM} = 2(m_f + 1)f_m = 2(\Delta f + f_m) \qquad (10.6)$$

When $m_f \ll 1$, it is narrow band modulation, and $B_{FM} \approx 2f_m$. When $m_f \gg 1$, it is wide band modulation, and $B_{FM} \approx 2\Delta f$. Δf is changed with m_f, in other words, the frequency structure and effective bandwidth will change with it, and the bandwidth will be broadened with m_f. The sideband is broadened with increasing m_f, thus the signal energy will distribute more evenly in the bandwidth. This makes the amplitude of fundamental frequency reduce sharply. The spectrums of sinusoidal carrier waveform and sine modulation waveform are shown in Figure 10.1.

The power spectrum density (PSD) defined signal power per unit frequency is an expression transformed from time-domain to frequency-domain, and it is just different expression forms of the same signal. PSD is studied from the point of view of energy and the relationship between the frequency spectrum and power spectrum is ultimately a relation of signal with power or energy. Figure 10.2 shows the comparison of PSD before and after sinusoidal frequency modulation, where the Welch method has been used (in this chapter, all PSD simulations are finished by Matlab@ and use the Welch spectrum estimation method). Here, $f_c = 50$ kHz, $f_m = 200$ Hz, $\Delta f = 12$ kHz, and $m_f = 60$. The result shows that the maximum value of the power spectrum envelope after modulation is significantly lower than the amplitude of non-modulation.

Changing frequency technology, which is also called spread spectrum frequency modulation (SSFM), means that its carrier frequency can be changed through some

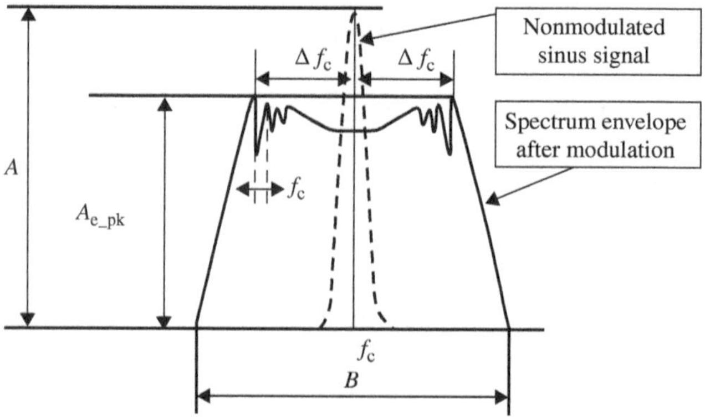

Figure 10.1 Spectrum of sine carrier wave after sine modulation [4]. © [2005] IEEE. Reprinted, with permission, from Balcells, J., Santolaria, A., Orlandi, A., et al. EMI reduction in switched power converters using frequency modulation techniques. *IEEE Trans. Electromagnetic Compatibility*, 2005

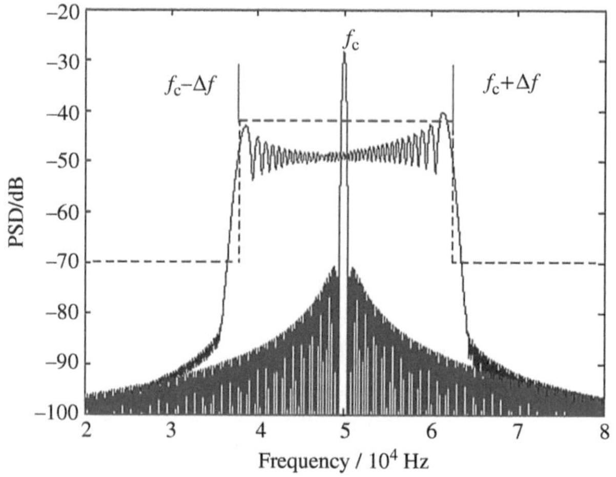

Figure 10.2 PSD of sine frequency modulation [2, 3]

modulation methods. This technology is first applied to deal with the radio frequency signal. By this technology, 98% of carrier wave power can be distributed in a particular area. As to switching converters, SSFM is just to make it work in a wider frequency range. Its basic idea is to spread the energy which is concentrated in the switching frequency and its harmonics to the surrounding frequency band by modulating the switching frequency of power devices and reducing the power spectrum amplitude on each frequency point. Over the past 10 years, there have been abundant research achievements in frequency modulating switching converters, mainly concluding periodic frequency modulation technology, jitter frequency, random frequency modulation, and chaotic frequency modulation. All these modulation technologies are aiming at the system clock as the target object, so they are also called spread spectrum clock generation (SSCG) techniques.

10.2.1.1 Periodic Frequency Modulation

Applying periodic frequency modulation to switching converters appeared in the middle of 1990s, and the method was first used on DC–DC converters by F. Lin of VPEC [5]. In this paper, the effects of the sinusoidal spread spectrum suppressing EMI for a forward DC–DC converter were studied. Then, Balcells *et al.* used periodic signals such as sine wave, triangular wave, and exponential function to modulate switching signals, and different modulation effects were compared [4].

10.2.1.2 Jitter Frequency Modulation

Jitter frequency is a special periodic frequency modulation technology which manifests with the carrier frequency having a periodic change within a certain range. It spreads the energy concerned in carrier frequency and its harmonics to the surrounding band, so that EMI on each frequency point can be reduced. Although this method cannot reduce the total interference's energy, the energy can be distributed in a wider band. Thus, it is easier to make EMI of switching power supply under the limits of EMI standards. Some companies such as Power Integration and ON Semi have developed various kinds of PWM control chip based on jitter frequency technology, such as TOPGX, NCP1230, and so on.

10.2.1.3 Random Frequency Modulation

Random frequency modulation means that switching frequency changes randomly with random distribution law in a certain frequency range, and the signal spectrum is transformed from the original spike like pulse to continuous distribution. So, a technically digital random signal producing circuit is needed using this method. The effects of random carrier frequency modulation suppressing the switching converter's conducted EMI have been studied by K.K. Tse. The effects of randomized pulse position, randomized pulse width, fixed duty cycle, and other modulation technologies suppressing EMI have been given in [6–8].

10.2.1.4 Chaotic Frequency Modulation

A chaotic signal has the characteristics of inherent randomness with the continuous spectrum of a deterministic system. Using a chaotic signal to modulate switching frequency can redistribute the energy in a wider band. Chaotic frequency modulation has the feature of constant duty cycle compared with other chaotic techniques (such as chaotic amplitude modulation, chaotic duty cycle modulation). It makes output voltage constant and has a small output ripple variation, little influence on other performances of the converter can be achieved. So it is an effective way to resolve the EMI problems of switching converters.

A comparison of different technologies of SSFM is shown in Table 10.1.

10.2.2 The Frequency Characteristics of Fixed Frequency PWM Wave

Consider that as a result of the shift and superposition of N unit pulses, fixed frequency PWMs are symmetrical to origin. Let its period be T, and duty cycle be d, that is

$$f(t) = \sum_{k=-\frac{N}{2}}^{\frac{N}{2}-1} f(t-kT) \tag{10.7}$$

The Fourier coefficient of the nth harmonic is

$$\dot{F}_n = \frac{2d}{T}\frac{\sin\left(\frac{n\omega d}{2}\right)}{\frac{n\omega d}{2}} = \frac{2d}{T}Sa\left(\frac{n\omega d}{2}\right) \tag{10.8}$$

From Equation 10.8, the fixed-frequency PWM has every order harmonics of a fundamental wave. The harmonic energy of a fixed-frequency PWM mainly focuses on a particular frequency and has a higher frequency spectrum peak. The PSD analysis using the Welch method of fixed-frequency PWM wave (f = 75 kHz, duty cycle d = 24%) is shown in Figure 10.3, and its power spectrum envelope line presents a clear S_a function curve.

Table 10.1 Performance comparison of different technologies of SSFM

Methods	Design complexity of hardware	Consequent on SPSP	Applications	Cost
Periodic frequency modulation	Medium to low	Medium	All kinds of switching converters	Medium to low
Jitter frequency	Medium	Medium	Specialized ICs	Medium–high
Random frequency modulation	Complex	Good	All kinds of switching converters	High
Chaotic frequency modulation	Medium	Good	All kinds of switching converters	Medium–high
Operating in chaotic state	Simple but related to systematic parameters directly	Good	Large ripple in low-frequency, limited in practice	Low, no need adding any other devices

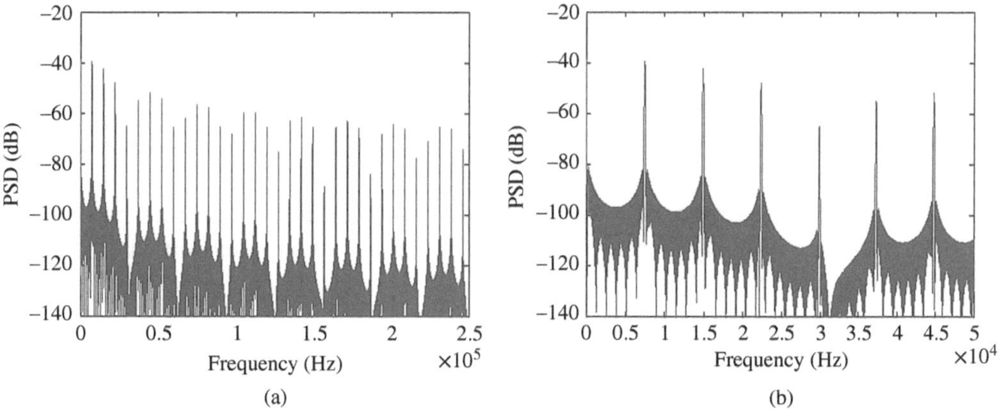

Figure 10.3 (a) PSD of fixed-frequency PWM signal and (b) enlargement of (a)

10.2.3 Frequency Characteristics of Spreading Frequency PWM Wave

There are harmonics with multipliers of the switching frequency of the PWM pulse, such as $2f_c$, $3f_c$, and so on [5]. Different harmonics have different modulation degrees m_{fn}, where n is the harmonic order of nonmodulating PWM waves, for example, $n = 2$ means second the harmonic of switch frequency, and $m_{fn} = nm_f$. Therefore, the spectrum results are different with the different order harmonics. For instance, if $m_f = 10$, then the modulation degree of the third harmonic m_{f3} will become 30. Applying the Carson bandwidth rule to every order harmonic, if $m_f \gg 1$, then we get

$$B_{n\mathrm{F}} = 2(nm_f + 1)f_m \approx nB_\mathrm{F} \tag{10.9}$$

The general trend of frequency modulation is shown in Figure 10.4. Here, every ordered harmonic power has been broadened. The higher the harmonic orders are, the

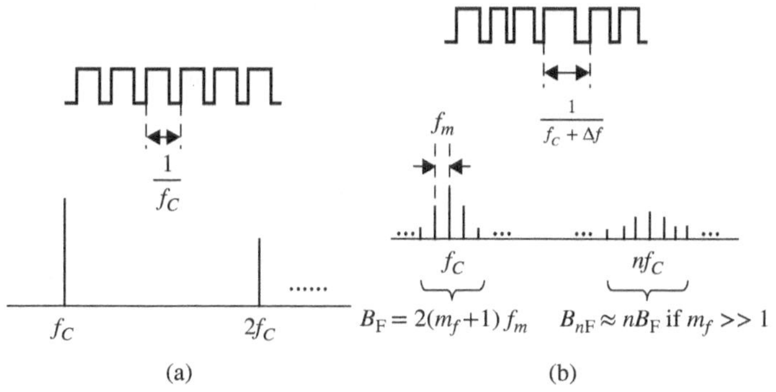

Figure 10.4 Schematics of PSD of PWM waveform for (a) square wave and (b) frequency modulation wave

wider the frequency spectrum band will be. With the increasing of harmonic orders, sideband aliasing may appear.

In general, the basic attributes of the spreading frequency PWM are as follows.

1. The results of harmonic suppression are enhanced with increments of modulation index m_f.
2. The results of the suppression harmonic are better for higher order harmonics.
3. There are no obviously increasing effects for harmonics suppression when Δf goes up to some extent.

The power spectrum of the sinusoidal modulation PWM wave is shown in Figure 10.5, where carrier $f_c = 75\,\text{kHz}$, duty cycle $d = 24\%$, modulation wave $f_m = 1.25\,\text{kHz}$, maximum frequency deviation $\Delta f_{max} = 7.5\,\text{kHz}$, modulation index $m_f = 6$, and modulation rate $\delta = 10\%$. It can be seen that the harmonics beside the switching frequency significantly come down, and the frequency band is broadened with the increasing of harmonic orders. The aliasing, which first appears at the ninth harmonic, causes new spectrum peaks to appear in several areas. In general, however, the power spectrum peaks are significantly lower than with un-modulation.

10.2.4 The Principle of Chaotic PWM Suppressing EMI

Provided the modulation signal is periodic, the power of the modulated signal will concentrate in a given frequency band thickly, and the spectrum peak will still be high. Therefore, one method to gain a lower EMI is by using a nonperiodic modulation signal, such as random or chaotic signals. In recent years, the bifurcation and chaotic behaviors of power electronic converters have been studied comprehensively, and using continuity characteristics of the chaotic spectrum to modulate power electronic converters' operating frequency to realize EMI suppression is widely covered. Paper [9] has pointed out that there are two advantages in using chaotic signals to the spreading spectrum instead of random signals, one is that it will never cause any performance loss, and the chaotic signal is easier to achieve than the random signal in engineering and its cost is lower. The other is that the chaotic signal has more controllable parameters such as invariant density, mixing rate, and so on. However, a random signal can only control a signal's PSD. So, chaotic modulation is more flexible than random modulation.

The most effective method of chaotic frequency modulation is to adjust the clock frequency of the PWM controller, as shown in Figure 10.6 [1, 10], where its chaotic signal can be produced by a digital processor or analog circuit. When a chaotic signal is produced by a digital controller and analog circuit, the holding time T $(=1/f_m)$ and the sampling interval T $(=1/f_m)$ are discrete and continuous respectively. In a practical power supply, the clock signal will be sent into the PWM controller after modulation and then produce chaotic modulation PWM signals to drive the switching semiconductor to work.

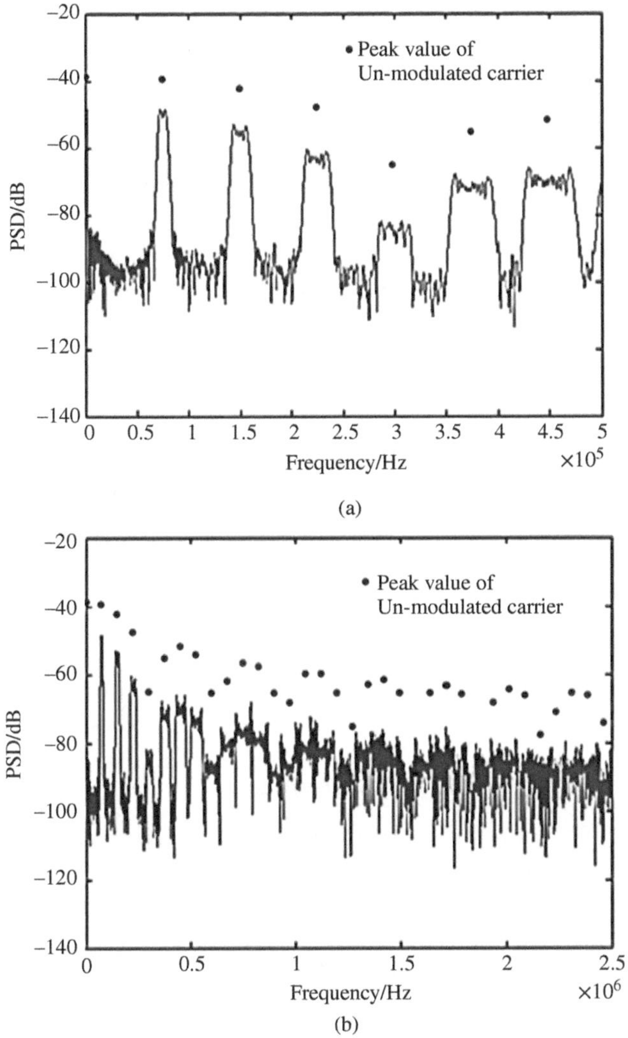

Figure 10.5 (a) PSD of PWM waveforms with sine modulation and (b) the enlargement of (a)

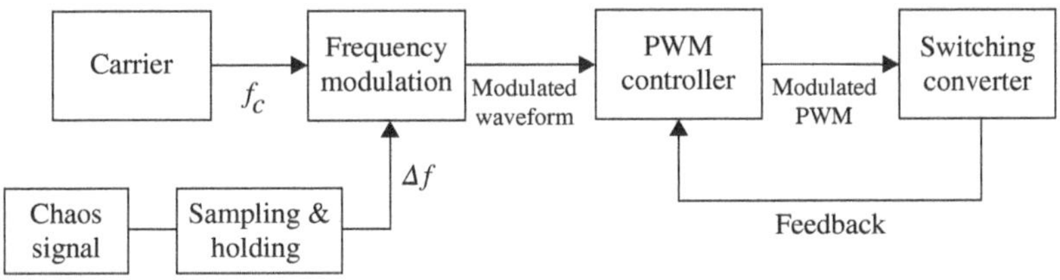

Figure 10.6 Block diagram of chaotic PWM modulation in power electronic converters

10.3 The Key Techniques of Chaotic PWM for Power Electronic Converters

The design method of chaotic PWM is the transfer from the fixed frequency PWM into chaotic PWM without changing the original characteristics of the power electronic converter too much. As to the engineering application of chaotic PWM modulation, the most likely and effective means is making fixed-frequency PWM become chaotic PWM based on the original PWM ICs. On the one hand, the system EMI may be suppressed by the continuous chaos spectrum. On the other hand, there is no need to change the raw circuit parameters and control circuit. So its rebuilding cost is lower, and the promotional value is remarkable. The principle of chaos suppressing EMI is shown in Figure 10.7. The charge–discharge time of the peripheral clock circuit of the PWM chip is modulated by the chaotic circuit, and then chaotic PWM waves with a certain modulation degree will be output. Therefore, there are two problems that should be resolved in engineering applications. One is how to control the related parameters of modulation, and the other is how to choose the right chaotic signal.

10.3.1 Parameter Selection of Chaotic PWM

According to Equation 10.6, there are modulation index m_f, maximum frequency deviation Δf, and modulation frequency f_{m} to influence PWM modulation bandwidth.

10.3.1.1 Maximum Frequency Deviation Δf

From the point of view of reducing EMI, a large Δf should be chosen because the band will be widened after SPSP, and the general distribution of harmonics will be flatter. The simulation result of chaos modulation by a nonlinear capacitor chaotic circuit (introduced in the next section) is shown in Figure 10.8. Here, the carrier $f_{\mathrm{c}} = 75$ kH, duty cycle $d = 24\%$, modulation rate of the real and dashed line are $\delta = 10$ and 20% respectively. This means that the maximum frequency deviations are $\Delta f = 7.5$ and 15 kHz respectively. The relationship between the maximum frequency deviation and bandwidth can be seen from Figure 10.8 clearly.

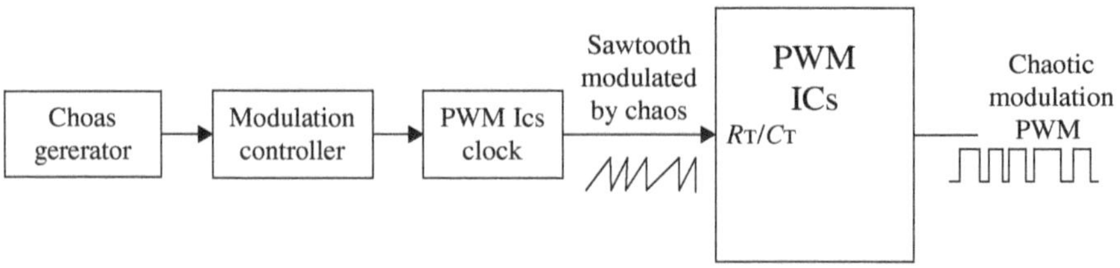

Figure 10.7 Schematic diagram of power electronic converters by chaos suppression EMI

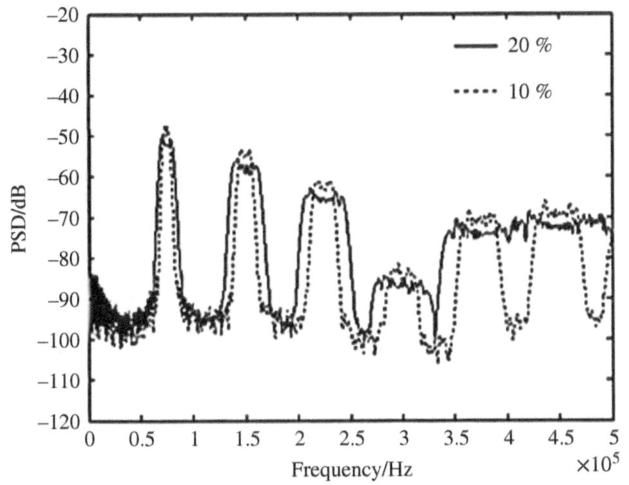

Figure 10.8 Influence of chaotic modulation bandwidth by the maximum modulation frequency

However, the maximum frequency deviation is not better for the bigger Δf. On the one hand, the larger Δf means that the possibility of aliasing in high order harmonics is larger, and the harmonics at some point will rise instead of reducing. On the other hand, from the point of view of the switching converter design, in order to make the overall system work properly, the frequency deviation value Δf should not be too large, and usually it meets $\Delta f < 10\% f_c$.

10.3.1.2 Modulation Index m_f

Based on frequency modulation theory, when $m_f \ll 1$, it belongs to narrow-band modulation (also called fast modulation). When the modulation index meets $m_f \gg 1$, it belongs to wide-band frequency modulation (also called slow modulation). Figure 10.9 is the simulation result of chaotic narrow-band modulation by logistic mapping, where its carrier $f_c = 75$ kHz, duty cycle $d = 24\%$, maximum frequency deviation $\Delta f = 7.5$ kHz, modulation rate $\delta = 10\%$, modulation index $m_f = 0.1$, and the mapping frequency is the same as the carrier. The frequency changing process is shown in Figure 10.9a, frequency between 7.1–7.9 kHz will change every 13.3 μs (1/75 kHz), Figures 10.9b,c are the comparisons of power spectrum before and after modulation. As to narrow-band modulation, although each harmonic becomes lower, its effect is not as remarkable as sinusoidal SPSP. Furthermore, the spectrum low frequency part is higher than the periodic state spectrum. So the system output ripple may be increased for this reason.

The simulation results of chaotic wide-band modulation using the population mapping method are shown in Figure 10.10. Where, $f_c = 75$ kHz, $f_m = 1.25$ kHz, $d = 24\%$, $m_f = 6$, $\Delta f = 7.5$ kHz, and $\delta = 10\%$. The frequency change process is shown

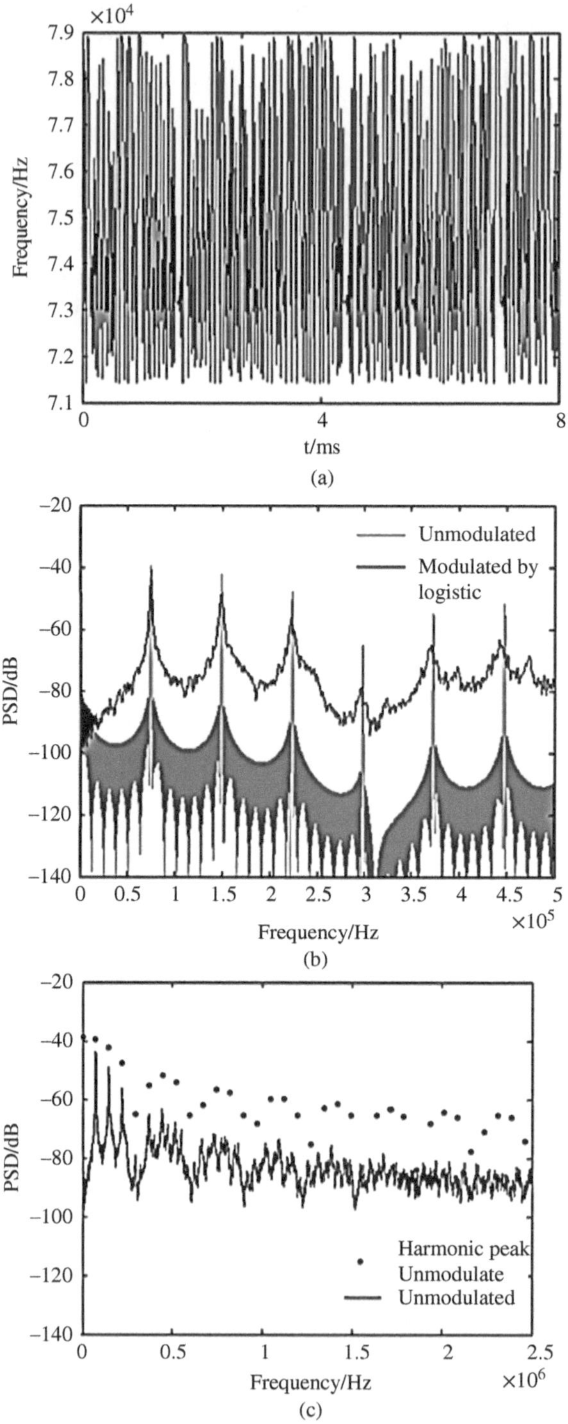

Figure 10.9 Narrow-band chaotic modulation by logistic map with (a) variation of frequency, (b) PSD (frequency: 0–500 kHz), and (c) PSD (frequency: 0–2.5 MHz)

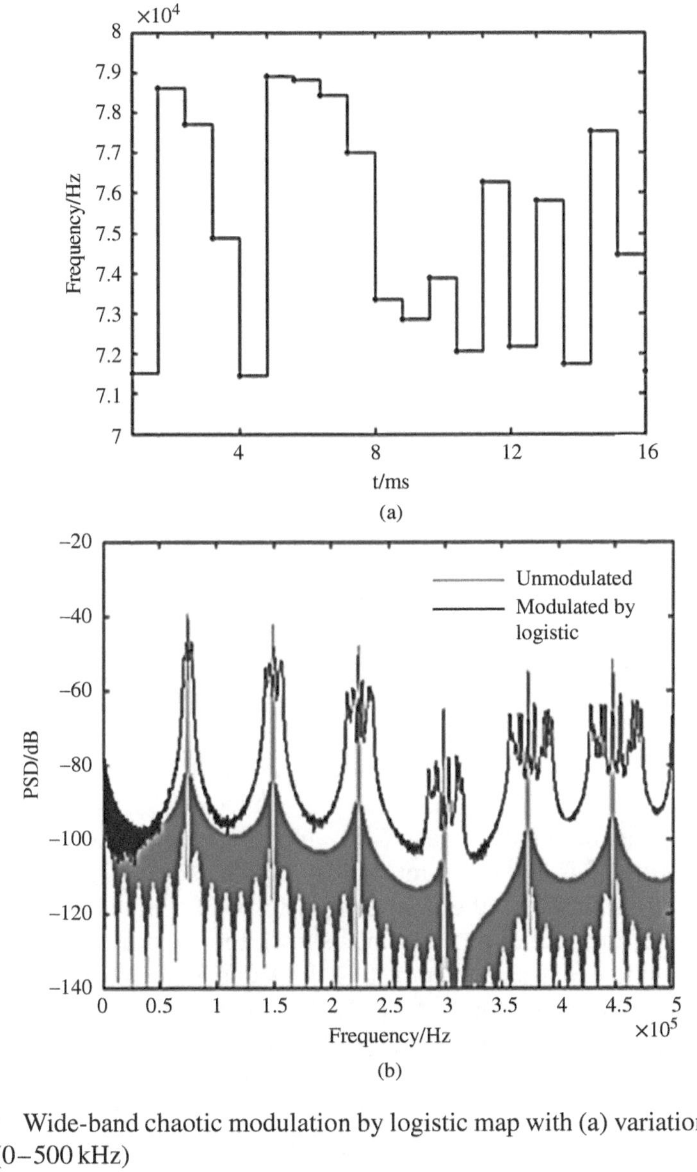

Figure 10.10 Wide-band chaotic modulation by logistic map with (a) variation of frequency and (b) PSD (0–500 kHz)

in Figure 10.10a, the PWM signal frequency changes every 0.8 ms (1/1.25 kHz). It is obvious that the harmonic band has been broadened in Figure 10.10b, this indicates that wide-band modulation is better than narrow-band modulation.

10.3.1.3 Modulation Frequency f_m

From the effects of modulation frequency influencing EMI, it needs to make f_m as small as possible to achieve a large enough modulation rate. But from the point of view of reducing the output filter sizes, it needs a higher f_m. Moreover, in order to gain the remarkable effects of suppressing harmonics, f_m should be larger than spectrum analyzer RBW. For the output ripples, f_m should be small enough such that output voltage fluctuation can be avoided, which means f_m should be significantly lower than the cross-over frequency of the control loop. Therefore, compromises need to be made based on the above points in practice.

10.3.2 Choice of a Chaotic PWM Modulation Signal

As for the switching converter, neither the carrier frequency nor maximum frequency deviation gain the freely changeable parameters of optimal EMC. In a practical circuit, the maximum frequency deviation is limited between 5–10% to avoid an irregular output wave.

It is generally known that the chaos phenomenon is unpredictable in the time-domain, but the distribution density is definite from a statistical point. Different chaotic signals have different probability densities, it is a new design angle to choose a proper chaotic modulation wave based on probability density being even or not. Paper [11] uses invariant density to deduce the theoretical calculation formula of every inductance current harmonic PSD on one order boost converter frequency. Paper [12] has shown that when the modulation rate is a little higher and chaotic sequence height is irrelevant, there is

$$\lim_{T\to\infty} \Phi_{\bar{s}\bar{s}}(f) = \frac{1}{2\Delta f}\bar{\rho}\left(\frac{f}{\Delta f}\right) \tag{10.10}$$

Here, $\Phi_{\bar{s}\bar{s}}(f)$ is the modulated mean and autocorrelation function after stripping carrier information, $\bar{\rho}$ is the invariant density of chaos mapping. According to the Wiener–Khintchine theorem, the Fourier transform of the autocorrelation function is the power spectrum. Equation 10.10 has shown the form of power spectrum after modulating by sinusoidal signal, and it is the same with the invariant distributed density of the modulation signal. Paper [12] also points out that, when the modulation index is large enough, the requirement of irrelevance is not too low and Equation 10.10 is also workable. Therefore, using chaotic mapping of evenly distributed density can make the power spectrum near every harmonic as flat as possible.

Generally speaking, there are two dominant ways to generate chaotic modulation. One is by numerical computation, the other is by a chaos oscillating circuit. Numerical computation is used to calculate the chaotic map continuously to get a group of discrete values. Usually, a special digital signal processor is needed to make it come true, but chaotic mapping and variable modulation rate can be flexible choices. General chaotic oscillating circuits include Chua's circuit, Van der Pol's oscillating circuit, and so on, which are all made of simple analog devices of small size and low cost. The frequency histograms of four chaotic systems – logistic mapping, the Henon map, Chua's circuit, and the nonlinear capacitor chaotic circuit are shown in Figure 10.11, where the sample number is evenly 10 000, the horizontal axis is the discrete value, and the vertical axis is the probabilities of every value.

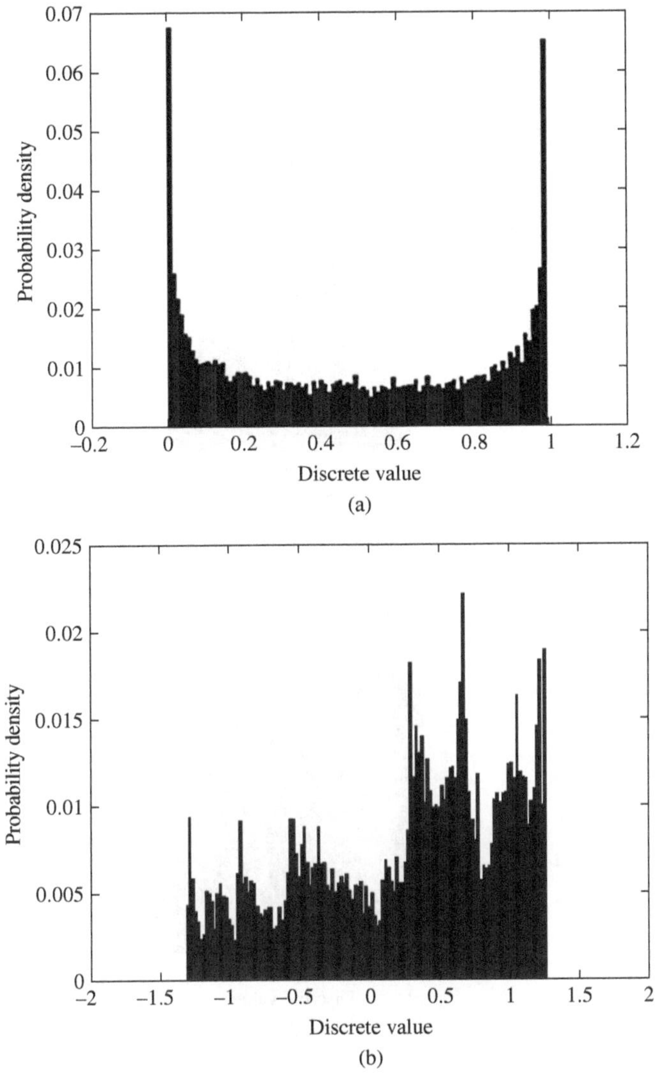

Figure 10.11 Probability histogram diagram for (a) logistic, (b) Rényi, (c) Chua's, and (d) PCCC

It can be seen that logistic mapping has the most uniform probability density, so it is the most suitable to be the chaotic modulation circuit, and the probability density of Chua's circuit is a triangular distribution, which is less uniform than the nonlinear capacitor chaotic circuit.

The simulation result of chaotic modulation using a nonlinear capacitor chaotic circuit is shown in Figure 10.12. Here, carrier frequency $f_c = 75$ kHz, duty cycle $d = 24\%$, maximum frequency deviation $\Delta f = 7.5$ kHz, modulation index $m_f = 6$ and modulation rate $\delta = 10\%$. As $m_f \gg 1$, the carrier power spectrum after modulation has the same

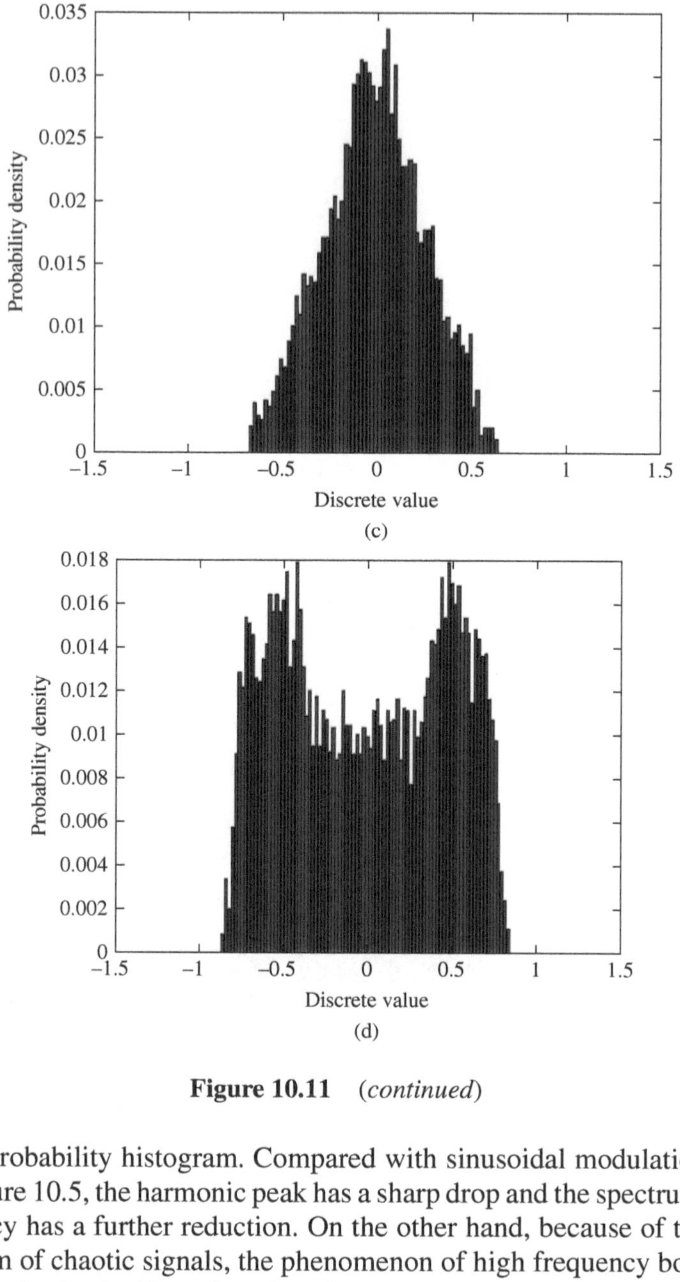

Figure 10.11 *(continued)*

shape as its probability histogram. Compared with sinusoidal modulation results, as shown in Figure 10.5, the harmonic peak has a sharp drop and the spectrum peak in the high frequency has a further reduction. On the other hand, because of the more uniform spectrum of chaotic signals, the phenomenon of high frequency boosting peaks caused by aliasing is significantly reduced.

10.4 Chaotic PWM Suppressing EMI Experiments

In this section, a new piecewise-linear capacitor chaotic circuit will be introduced first, and then this circuit producing chaotic modulation signals is used to realize chaotic modulation of DC–DC converters and DC–AC converters.

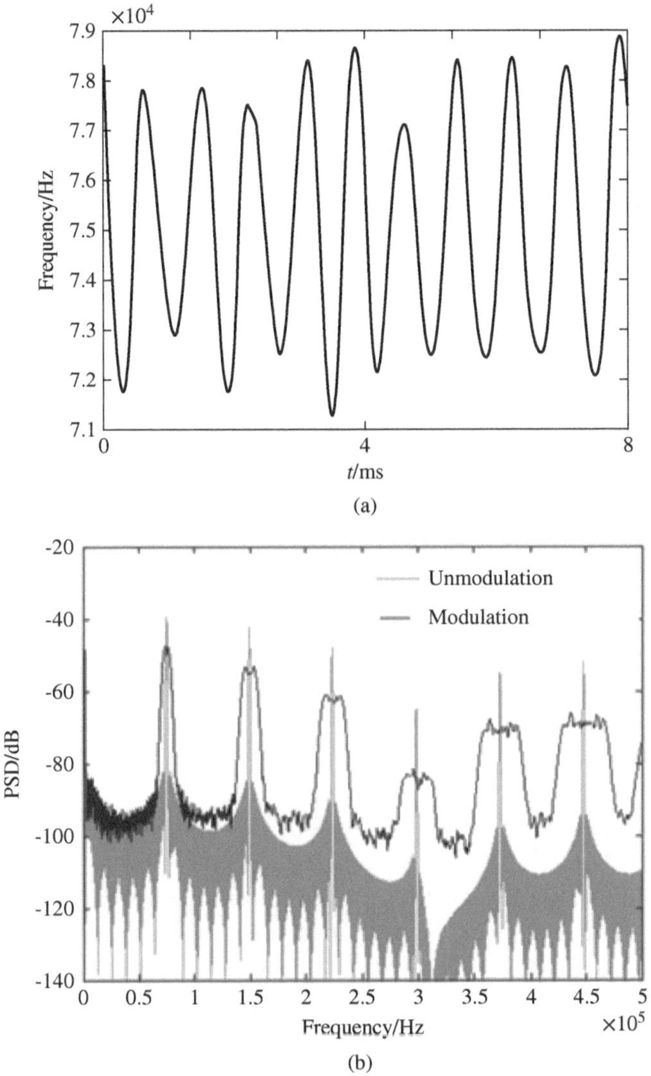

Figure 10.12 PSD of chaotic modulation PWM waveforms with (a) variation of frequency, (b) PSD (0–500 kHz), and (c) PSD (0–2.5 MHz)

10.4.1 Modulation Circuit of Piecewise-Linear Capacitor Chaos Circuit

Chua's circuit, one of the most famous chaos circuits, has a piecewise-linear negative resistor and exhibits a wide variety of nonlinear dynamics phenomena such as bifurcation and chaos. Gao *et al.* [13] introduced a new chaos circuit whose core component is a piecewise-linear capacitor. The new chaos circuit is a compact three dimensional autonomous circuit whose component quantity is the same as Chua's chaos circuit

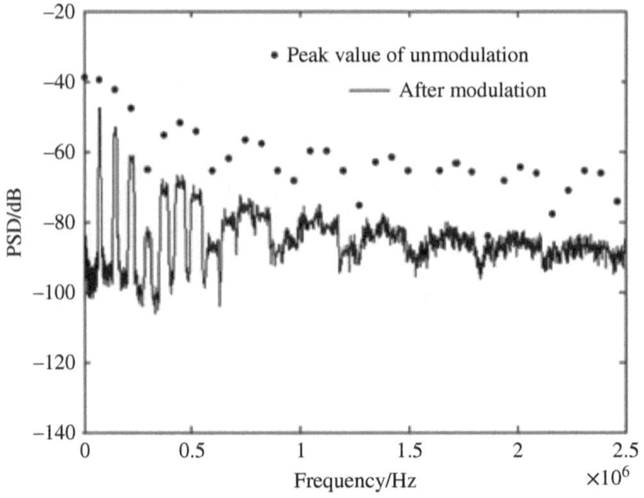

(c)

Figure 10.12 *(continued)*

but not topologically conjugated. This circuit is better than Chua's circuit because the power the piecewise-linear capacitor consumes is lower [12].

We define a piecewise-linear charge-controlled capacitor which has the following q-v characteristics

$$v_{C1} = f(q) = \frac{1}{C_b}q + \frac{1}{2}\left(\frac{1}{C_b} - \frac{1}{C_a}\right)(|q - q_0| - |q + q_0|) \tag{10.11}$$

The chaos circuit employing this piecewise-linear capacitor C_1 consists of one inductor L, one linear resistor G, one linear negative resistor $-G_0$, and one linear capacitor C_2, as shown in Figure 10.13.

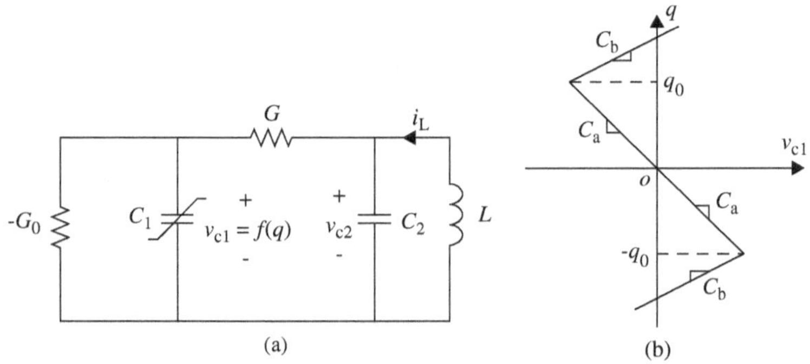

Figure 10.13 (a) Schematic diagram of piecewise-linear capacitor chaos circuit and (b) q-v characteristics of the piecewise-linear capacitor [15]

The experimental circuit is illustrated in Figure 10.14, where the linear negative resistor is composed of OA_1 and R_0–R_2, the piecewise-linear capacitor is composed of OA_2, C_0, and R_3–R_4. The parameters are shown in Table 10.2. All operational amplifiers are LF353 with supply voltage at ±15 V. In Figure 10.15, the equivalent inductor L is made of an impedance transformer [14], so that this chaos circuit does not rely on the use of inductors. Only capacitors, resistors, and operational amplifiers are employed.

Using the parameters as shown in Table 10.2, the negative resistor $-G_0 = -1/(2R_0) = -0.05\,\text{mS}$, the equivalent inductor $L = 0.32\,\text{H}$, $C_a = -6.6\,\text{nF}$, $C_b = 4.4\,\text{nF}$ and the piecewise-linear capacitor has a form as Equation 10.11.

It is difficult to measure the charge of the piecewise-linear capacitor directly. But we could measure the charge q through measuring v_{c0}. Because $i_3 << i_1$ and i_2, we can regard $i_1 = i_2$ approximately. As well as the charge of the piecewise-linear capacitor $q = \int i_1 dt$ and $i_2 = C_0 \frac{dv_{c0}}{dt}$, we can get $q = C_0 v_{c0}$. Thus, the voltage of capacitor C_0 is in direct proportion with the charge q. There is a proportional coefficient between them, obviously.

An experimental waveform of a single scroll chaotic attractor is shown in Figure 10.16a with $R = 10\,\text{k}\Omega$. Referring to normalization forms introduced earlier, the computed dimensionless parameters based on the above components are $\alpha = 6.8$, $\beta = 14$, $\gamma = 0.5$, $a = -1$, and $b = 1.5$. Also there is an experimental waveform of a

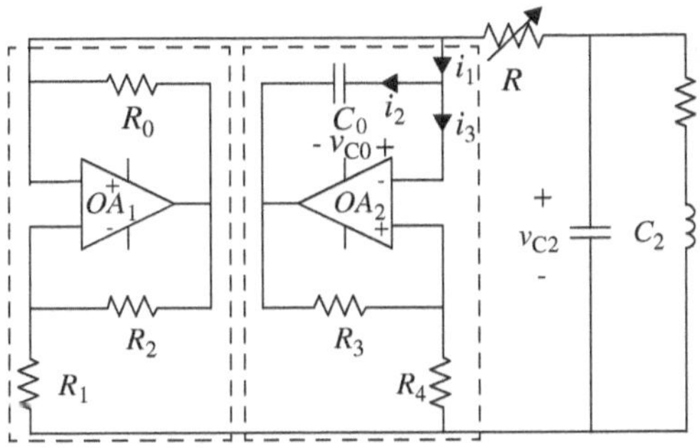

Figure 10.14 Experimental circuit scheme of piecewise-linear capacitor chaos circuit [16]

Table 10.2 Parameters of the piecewise-linear capacitor chaos circuit

Circuit components	Values	Circuit components	Values
R (adjustable)	20 kΩ	C_2	45 nF
R_0	10 kΩ	R_1	20 kΩ
R_2	10 kΩ	R_3	30 kΩ
R_4	20 kΩ	C_0	4.4 nF
R_L	10 Ω	R_5–R_7	1 kΩ
R_8	1.45 kΩ	C_3	0.22 μF

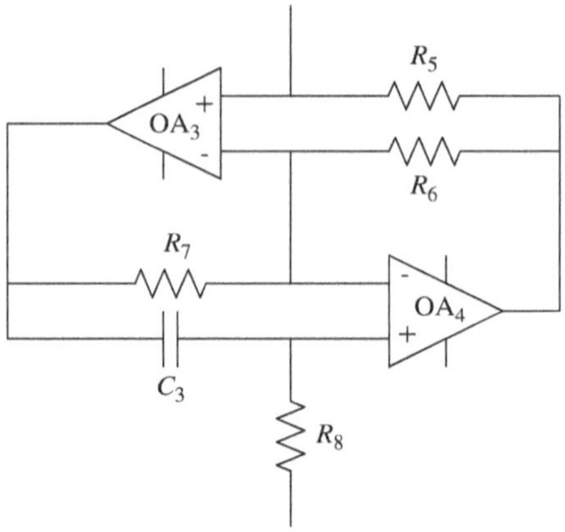

Figure 10.15 Operational amplifier realization of simulated inductor [16]

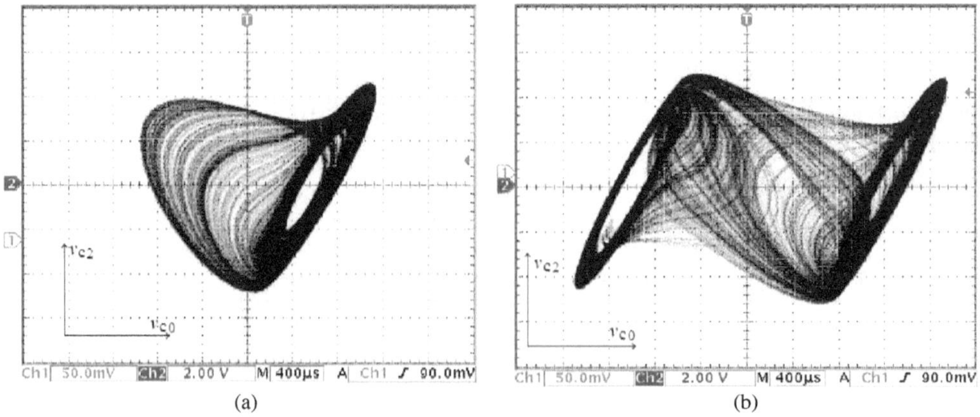

Figure 10.16 Experimental phase portraits of chaos attractors for (a) single scroll chaos attractor and (b) double scroll chaos attractor

double scroll chaotic attractor in Figure 10.16b with $R = 8\,\text{k}\Omega$, whose dimensionless parameters are $\alpha = 6.8$, $\beta = 9$, $\gamma = 0.4$, $a = -1$, and $b = 1.5$. Experiments prove that the piecewise-linear capacitor is feasible, and this chaos circuit has rich chaotic dynamic motions.

Experimental time-domain waveforms and FFT analysis of double scroll outputs are shown in Figure 10.17, where its center frequency is 1.25 kHz.

10.4.2 The DC–DC Converter Suppressing EMI Based on UC3842

UC3842 is a current controlled pulse width controller, its main principle is to make the output inductance current peak variable with error voltage by regulating the duty cycle, and it is a popular PWM control chip applied to all kinds of commercial power

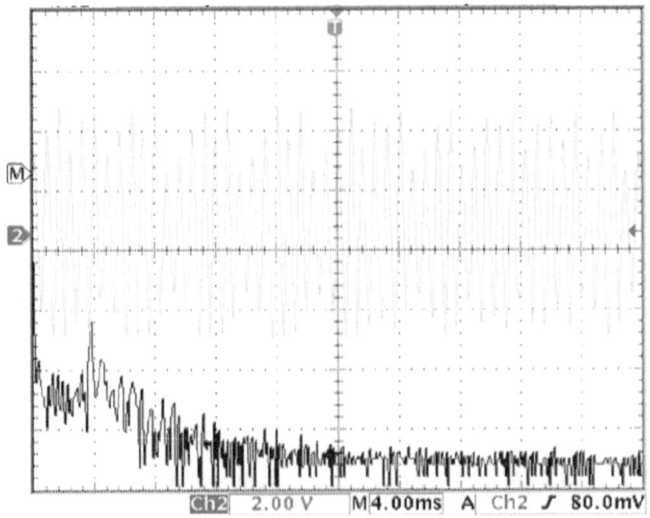

Figure 10.17 Experimental waveforms of double scroll outputs (top) and FFT (below)

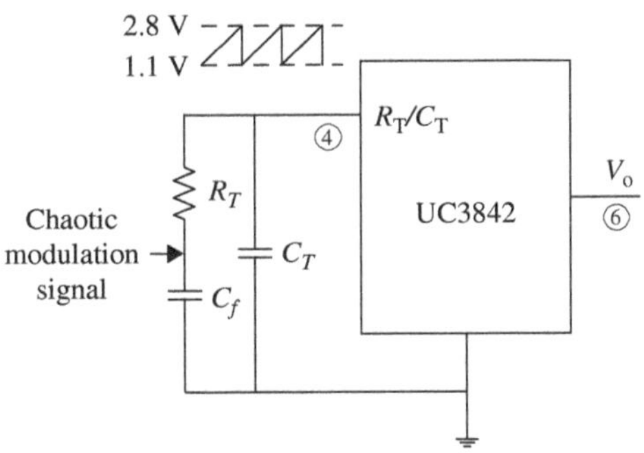

Figure 10.18 Schematic diagram of chaotic modulation UC3842

supply products. Therefore, it is acceptable, single, and low cost to make chaotic suppressing EMI rebuilding on this chip.

The principle of chaotic modulation PWM controlled chip UC3842 is shown in Figure 10.18. The methods are as follows: the chaotic signal produced by the piecewise-linear capacitive chaotic circuit from v_{c2} to the area between R_T and C_f of UC3842 makes the chip inner oscillating circuit charge current change with the amplitude of the modulation signal, so that it can lead charge time changing. The variation of the sawtooth wave frequency with chaotic modulation signal amplitude will make the output PWM driving signal frequency change as well. In brief, this is a chaotic frequency modulation circuit which uses chaotic signal amplitude to regulate PWM wave frequency.

Use the PCCC circuit as the chaotic signal generator circuit, and make the capacitor C_f (shown above) voltage hover around 5 V as chaotic regulation through a proper

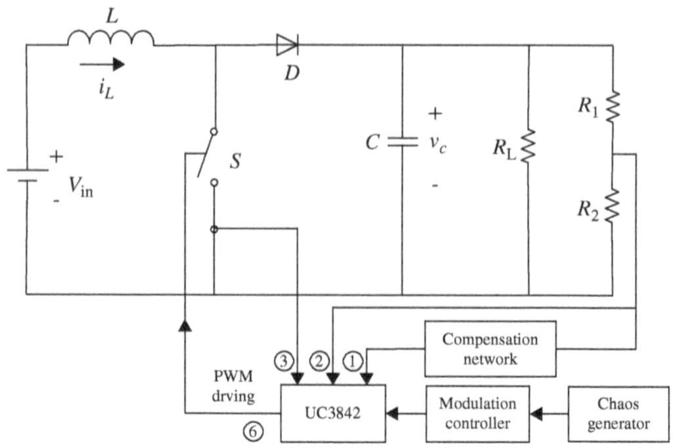

Figure 10.19 Schematic diagram of chaotic modulation boost converter [16]

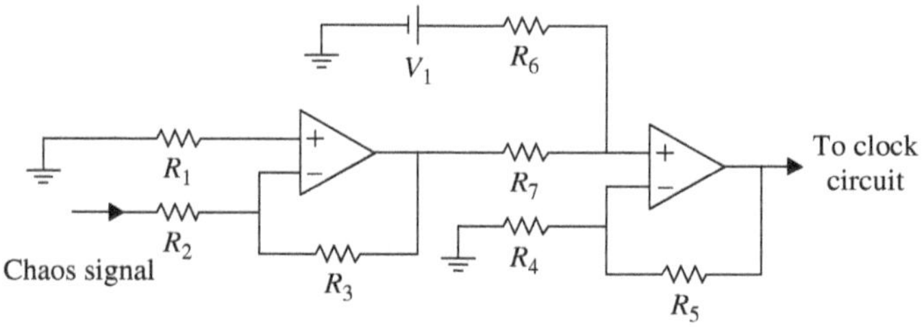

Figure 10.20 Control circuit of modulation depth [16]

modulation degree controlled circuit. Applying this circuit to a boost converter, its schematic diagram is shown in Figure 10.19. The modulation depth controlled circuit is shown in Figure 10.20, and this circuit can make sampling to chaotic signals and regulate maximum frequency deviation.

10.4.2.1 Simulation Research

Software Simetrix based on UC3842 SPSP modulation has been used to construct the simulation circuit, it main parameters are $E = 10\,\text{V}$, $L = 500\,\mu\text{H}$, $C = 220\,\mu\text{F}$, $R_L = 20\,\Omega$, $R_1 = 20\,\text{k}\Omega$, $R_2 = 4\,\text{k}\Omega$, the switching frequency is 75 kHz and output voltage is 14 V. Figure 10.21 shows UC3842 output PWM driving signal FFT under sinusoidal modulation and chaotic modulation with the same maximum frequency deviation $\Delta f = 7.5\,\text{kHz}$.

Referencing these three kinds of frequency modulation effects, the spectrum can become continuous and its harmonic peak drops significantly by using both the periodical and chaotic modulation technique. But, in general, the chaotic modulation effect is better. In order to analyze the EMI suppressing effect by different spread spectrum technologies further, the decreasing situation of harmonics peak under 10th orders

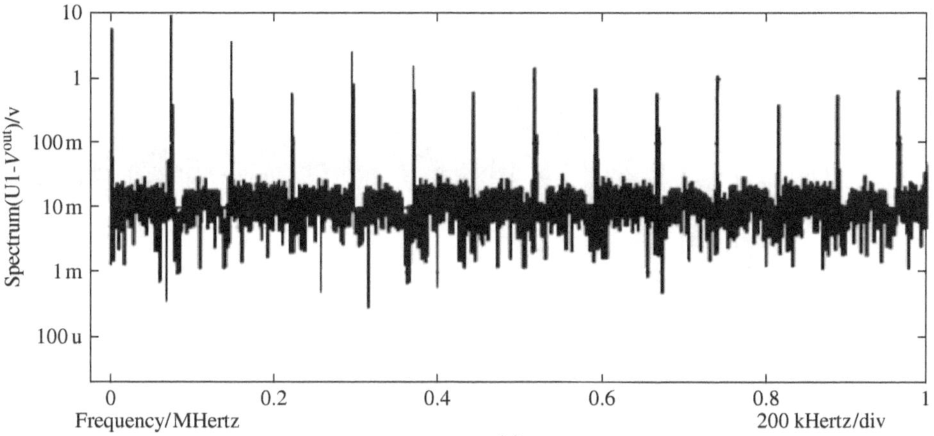

(a)

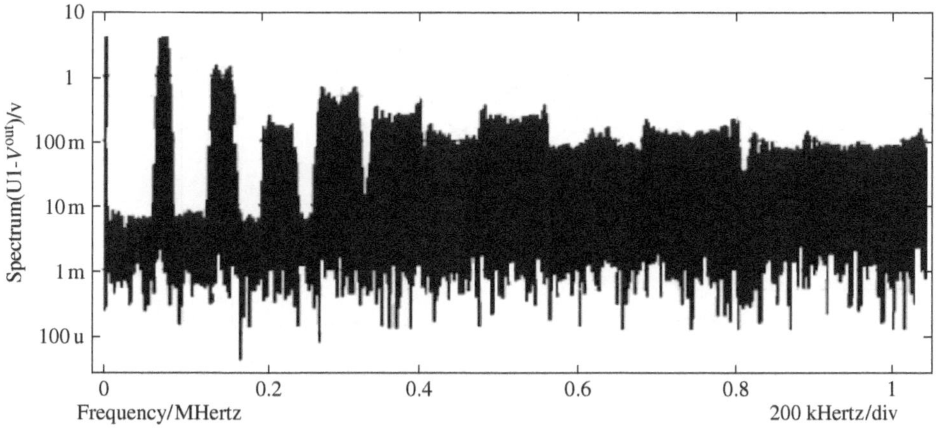

(b)

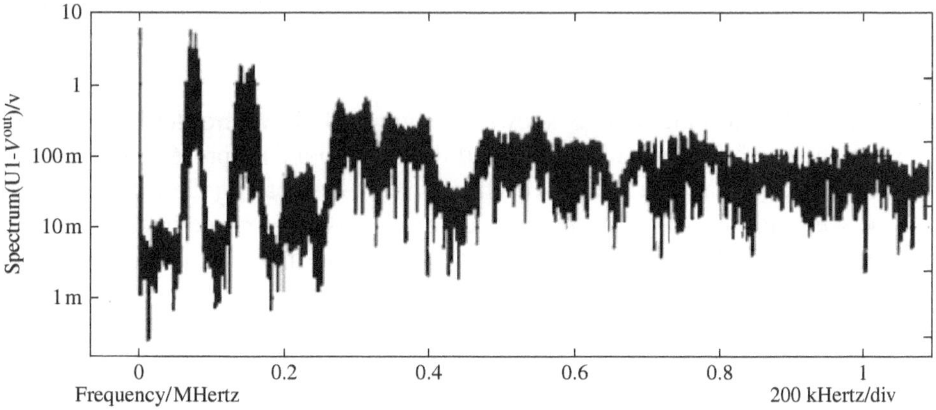

(c)

Figure 10.21 Simulation results of drive signal of UC3842 by frequency modulation with (a) un-modulated, (b) sine-modulated, (c) Chua's circuit, and (d) PCCC

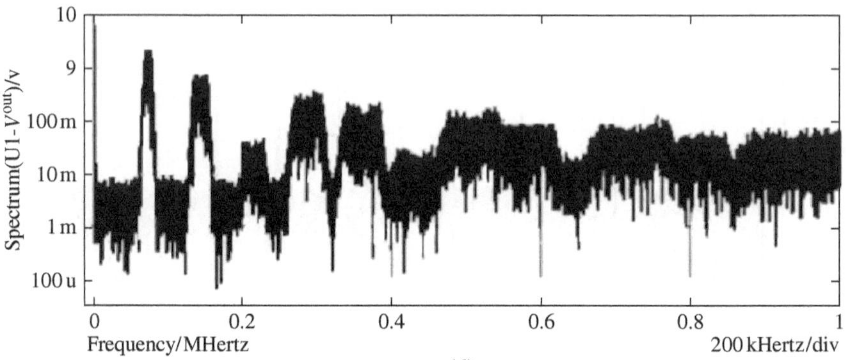

(d)

Figure 10.21 *(continued)*

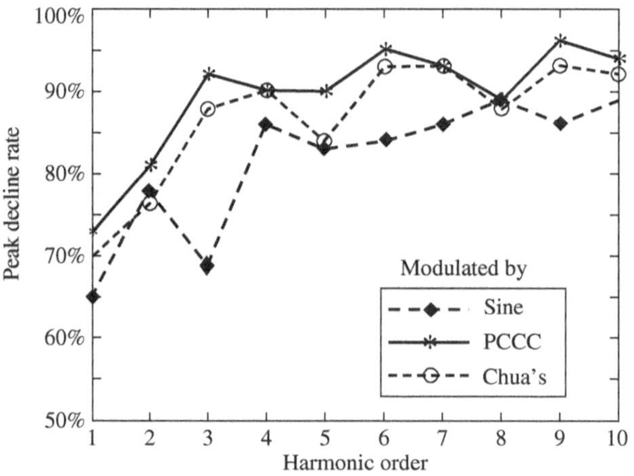

Figure 10.22 Comparison of the decreasing amplitude of the harmonic peak under different frequency modulation methods

with three modulation methods is shown in Figure 10.22, where the horizontal axis represents the harmonic order number, and the vertical axis represents the harmonic peak ratio before and after modulation (linear axis). Comparing the different results, it can be seen that the spectrum peak value of all modulation methods drops as the harmonic order number increases. Although sinusoidal modulation can broaden the frequency band and reduce the harmonic peak spectrum, as the chaotic modulation harmonic peak can be lower, and spectrum is more even, its modulation effect is better.

10.4.2.2 Experiment Result

The experimental waveform of a chaotic modulation boost converter is shown in Figure 10.23. From the driving signal spectrum after modulation, it can be seen

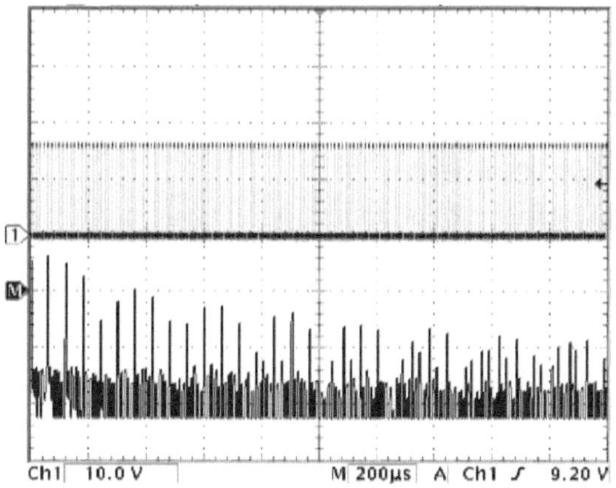

(a)

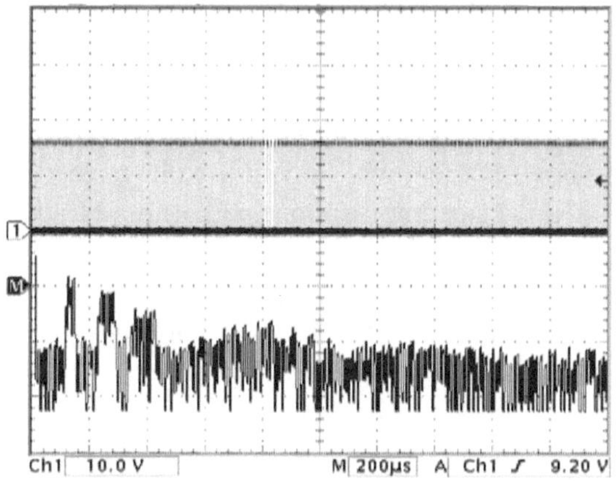

(b)

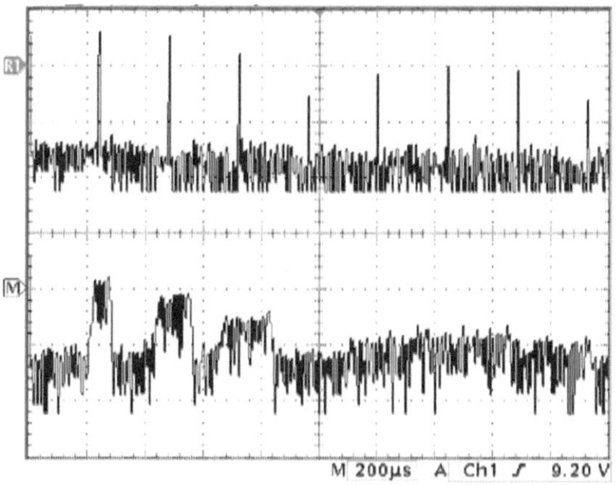

(c)

Figure 10.23 Experimental waveforms of a power semiconductor. Here, (a) is the drive signal (top) and FFT (below) before modulation, (b) is the drive signal (top) and FFT (below) after modulation, and (c) is the comparison FFT signal of original (top) and chaotic (below) modulation that each switching frequency harmonic has been broadened and harmonic peaks are reduced in some degree after chaotic frequency modulation. Moreover, in the high order harmonics, the harmonic peaks almost disappear. After modulation, the switching frequency harmonic peak has been reduced over 8 dB, and even drops by over 15 dB in some higher harmonics.

The experimental waveforms of chaotic modulation output voltage ripple are shown in Figure 10.24. The output voltage ripple has increased from about 0.3 to 0.5 V after modulation, the ripple has not changed so much after chaotic modulation, and its output voltage fluctuation is within the allowable range. It can be seen from the experiment results that the ripples slightly rise as the modulation degree increases.

10.4.3 EMI Suppression of Full Bridge Inversion Based on SG3525

SG3525 is a current controlled PWM controller IC. The inductance current and error amplifier output signal can be directly compared with the input terminals of a PWM comparator, so that the duty cycle can be regulated to make out the inductance peak current change with error voltage. Because there are both voltage loops and current loops in the circuit, it shows some improvement in voltage regulation of switch power supply, load regulation, and transient response performance, so it is an ideal new controller at present.

The scheme of chaotic frequency modulated PWM controller IC SG3525 is shown in Figure 10.25. Enter the chaotic modulation signal into terminal R_T of SG3525 directly to make the chip inner oscillation circuit charge current change

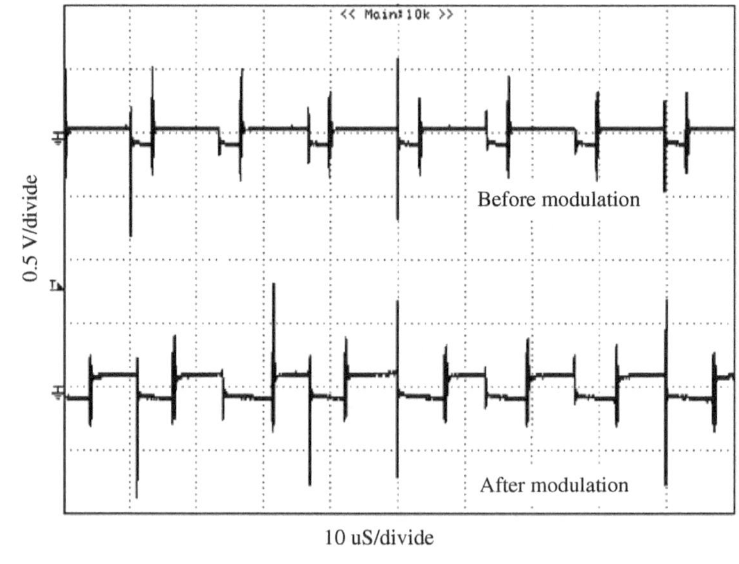

Figure 10.24 Experimental waveforms of output voltage

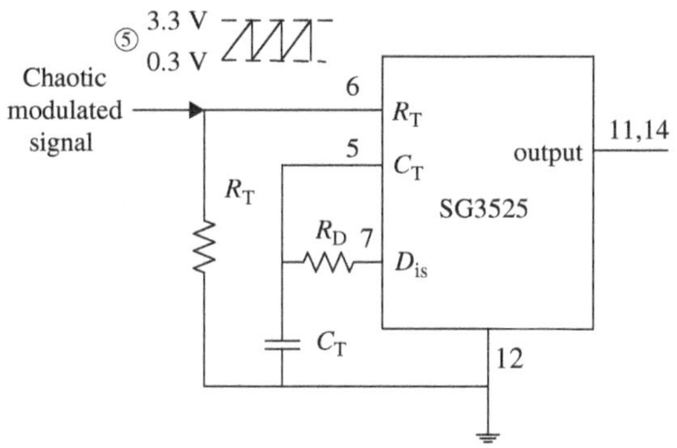

Figure 10.25 Schematic diagram of chaotic modulation SG3525

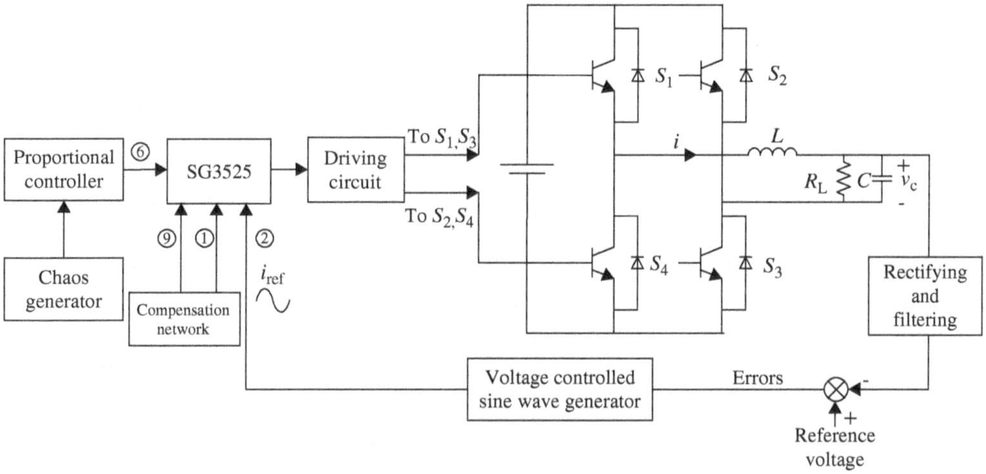

Figure 10.26 Schematic diagram of chaotic modulation of full bridge inverter [16]

with modulation signal amplitude, so that the charge time can be changed, and then sawtooth wave (STW) frequency will change with the amplitude of the chaotic modulation signal and the frequency of the PWM output will swing as well.

The experimental circuit of a full bridge sinusoidal inverter is shown in Figure 10.26, its main parameters are $f_c = 20\,\text{kHz}$, $E = 10\,\text{V}$, $R = 1\,\Omega$, $L = 0.5\,\text{mH}$, and $C = 220\,\mu\text{F}$. It uses MPT50N06 as a switching tube and the drive circuit uses TLP250.The output voltage average value of the inverter after transformer isolation and rectifier filter can be achieved. Comparing this value with reference voltage will generate an error signal. A level controlled sinusoidal wave generator is used to produce a sinusoidal wave whose amplitude changes with error signals. After comparing the boosting sinusoidal wave($1-3.2\,\text{V}$) with the chaotic modulation STW signal by SG3525, the chaotic modulation SPWM drive signal can be received, and then the operating state of the four switching tubes in the full bridge circuit can be controlled.

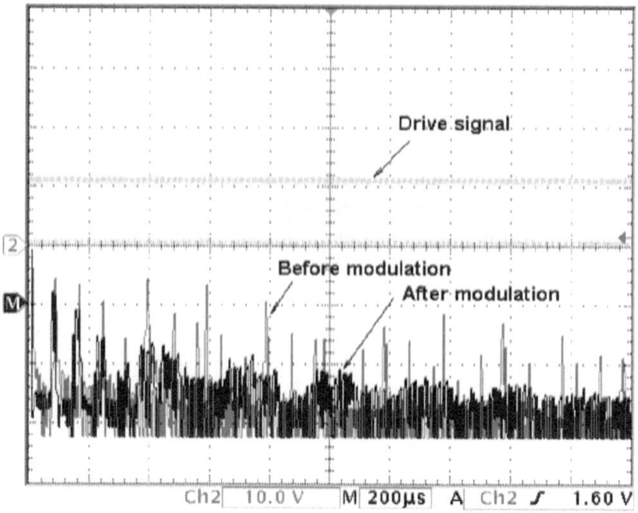

Figure 10.27 Experimental waveforms of drive signal (top) and FFT (below)

The experimental wave shape using the chaotic modulation sinusoidal pulse width modulation (SPWM) drive signal is shown in Figure 10.27. Each harmonic peak has decreased after modulation, the spectrum frequency becomes obviously broadened and continuous, and the higher order harmonics are the more significant peak declines. The inverter inductance current waves after chaotic modulation are shown in Figure 10.28. Before modulation, the switching frequency is fixed, and the harmonics are focused on the switching frequency integer multiples area. After modulation, the current wave period becomes jittering and the switching frequency is no longer constant, the spectrum has been broadened and every harmonic peak has dropped with varying degrees. It can be seen from the current waveforms and power frequency harmonics before and after modulation in Figure 10.29, that chaotic modulation does not have too much influence on the inverter output waveforms. And it is almost unchanged for power frequency harmonics after modulation, so chaotic modulation is an effective way to reduce sinusoidal inverter EMI.

10.5 EMI Suppression of Commercial Switching Power Supply

A chaotic frequency modulation experiment is performed in an AD–DC power supply (PART NO.418-000XXXXX) developed for an IBM desk computer by Emerson (ASTEC) company. The experiment principle circuit and parameters are illustrated in Figure 10.30 and Table 10.3. The PWM control IC is SG6846 produced by the System General Company and has wide commercial application. The sine frequency modulation and chaotic frequency modulation are tested in this prototype, where the

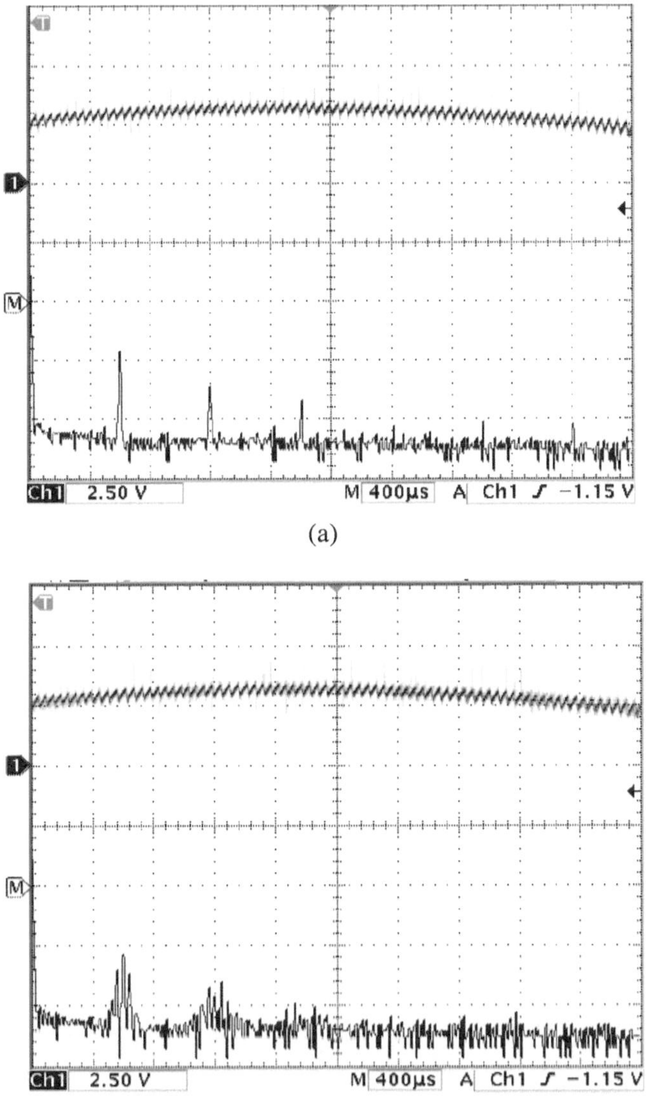

(a)

(b)

Figure 10.28 Experimental waveforms of inductor current (top) and its FFT (below) (a) before modulation and (b) after modulation

sine modulation signal is obtained by the AC source and the chaotic modulation signal is obtained by Chua's circuit.

All the experimental waveforms are captured in the operation of full load as shown in Figure 10.31. We may see: (i) the spectrum shapes of the PWM driver by sine spreading and fixed frequency are almost same as the discrete spectrum; (ii) the

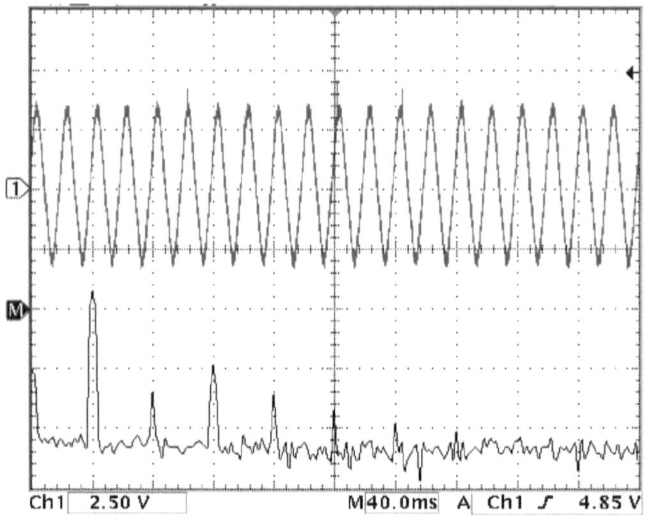

(a)

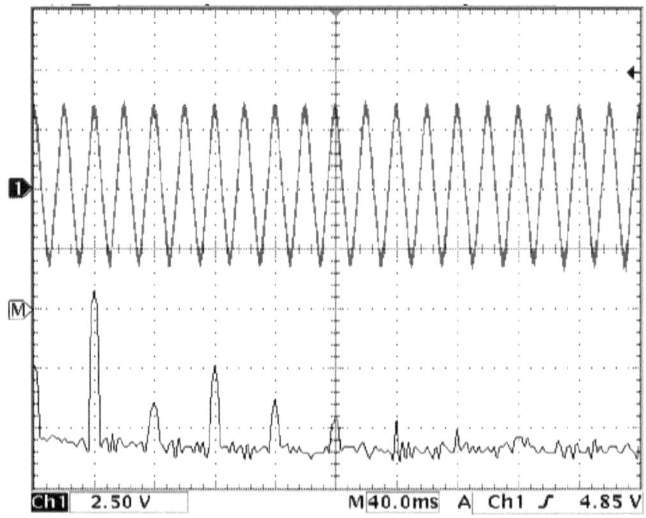

(b)

Figure 10.29 Experimental waveforms of inductor current (top) and its FFT (below) (a) before modulation and (b) after modulation

spectrum peak value of sine spreading is only 1–2 dB less than the fixed frequency; and (iii) the spectrum of chaos spreading continues, and there is a 10 dB drop at several spectrum peaks.

For the different frequency modulation methods, the ripples of output voltage are shown in Figure 10.32. We may find that the frequency of the output voltage by chaos

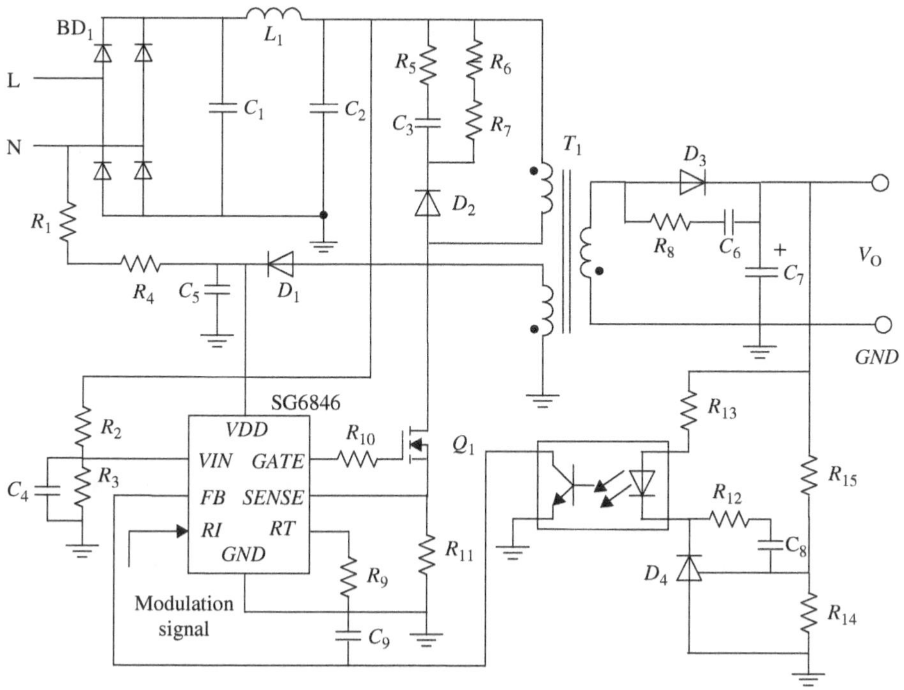

Figure 10.30 Experiment principle circuit [17]

Table 10.3 Experiment parameters

Power	Input	Output	Frequency	Δf
65 W	220 V AC	12 V5A	100 kHz	10 kHz

modulation is changed via the spreading signal, but the ripples' peak and shape are essentially unchanged. Therefore, it is feasible to use the chaos modulation technique to lower EMI, and perhaps this is the future development direction of the switching power supply.

The conducted EMI test of the experimental prototype was held in the shielded room of Rohde & Schwarz@, where the EMI test frequency range is from 150 kHz to 30 MHz and the standard is CISPR PUB.22 CLASS B. According to the power supply industry request, there would be 6 dB margin compared to the limitation of CISPR. In Figure 10.33 [17], our test results show that the harmonic energy distributing beside frequency nf_s caused the QP and AV limit excess and the EMI test failed. After sine modulation, the harmonic energy is widened and the peak value is lowered which causes the AV test value to be less than the AV limit. But the QP test value is still excessive at 3 MHz. After chaos modulation, the energy broadens further. In the low

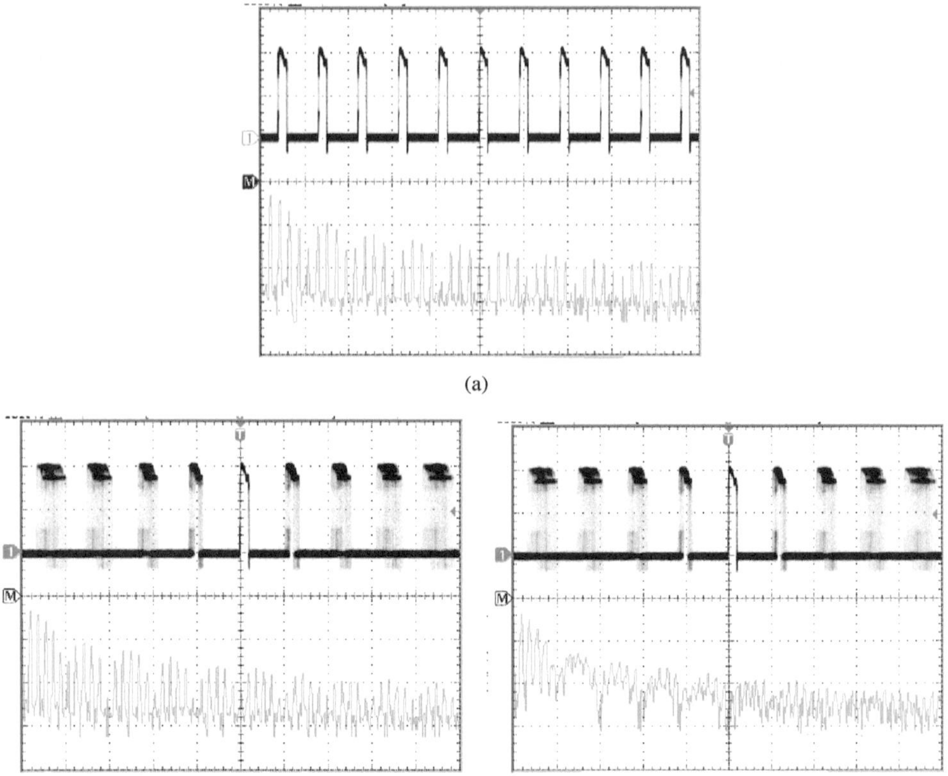

(a)

(b) (c)

Figure 10.31 Experiment waveforms of PWM driver (up, x axis: 10 μS/divider, y axis: 10 V/divider) and FFT spectrum (down, x axis: 500 kHz/divider, y axis: 20 dB/divider) with (a) fixed-frequency PWM, (b) spread spectrum by sine signal, and (c) spread spectrum by chaos signal

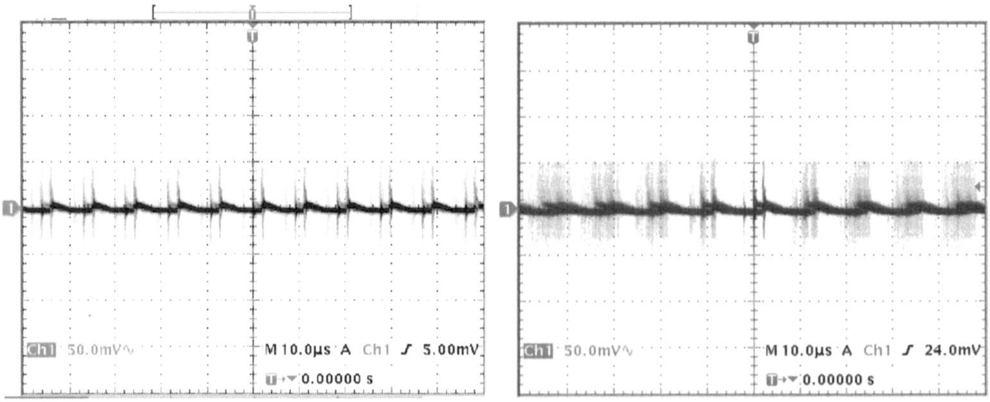

(a) (b)

Figure 10.32 Ripples of output voltage with (a) fixed-frequency PWM and (b) chaotic frequency modulation

CONDUCTED EMI 31 Oct 2005 11:16

CISPR PUB.22 CLASS B LESS 6dB

EUT:
Manuf: ASTEC--CDE
Op Cond: 230VAC
Operator:
Test Spec:
Comment: L1 and N tested

Scan Settings (2 Ranges)

Frequencies			Receiver Settings					
Start	Stop	Step	IF BW	Detector	M-Time	Atten	Preamp	OpRge
150kHz	3MHz	0.8%	10kHz	PK+AV	50msec	Auto	OFF	60dB
3MHz	30MHz	0.4%	10kHz	PK+AV	5msec	Auto	OFF	60dB

Transducer	No.	Start	Stop	Name
	11	9kHz	30MHz	LocalMea

Final Measurement: Detectors: X QP / + AV
Meas Time: 1sec
Subranges: 50
Acc Margin: 6dB

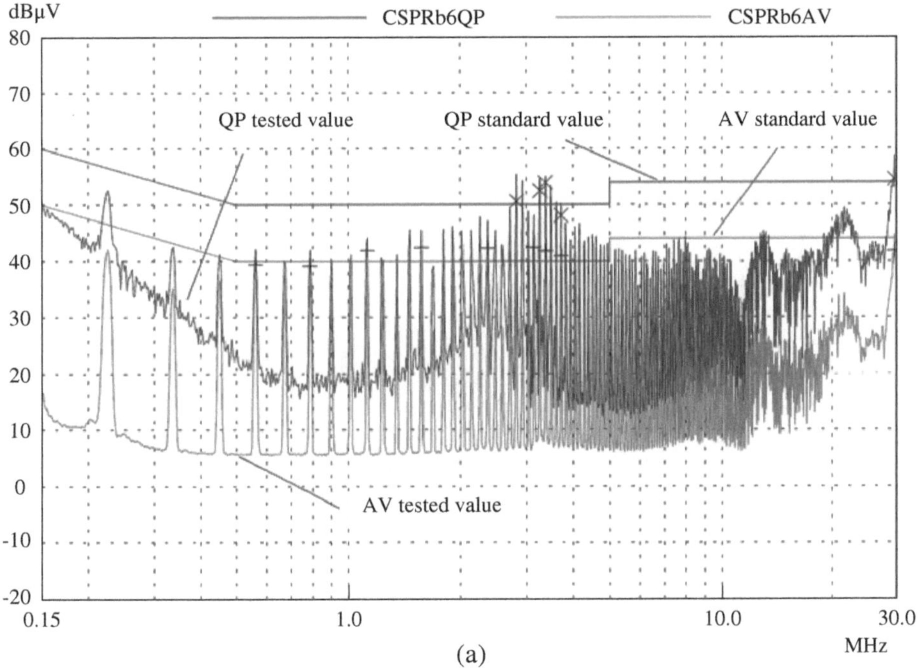

(a)

Figure 10.33 Conducted EMI test results with (a) fixed frequency PWM, (b) sine modulation, and (c) chaos modulation

CONDUCTED EMI 31 Oct 2005 14:19

CISPR PUB.22 CLASS B LESS 6dB

EUT:
Manuf: ASTEC--CDE
Op Cond: 230VAC
Operator:
Test Spec:
Comment: L1 and N tested

Scan Settings (2 Ranges)

Frequencies			Receiver Settings						
Start	Stop	Step	IF BW	Detector	M-Time	Atten	Preamp	OpRge
150kHz	3MHz	0.8%	10kHz	PK+AV	50msec	Auto	OFF	60dB
3MHz	30MHz	0.4%	10kHz	PK+AV	5msec	Auto	OFF	60dB

Transducer	No.	Start	Stop	Name
	11	9kHz	30MHz	LocalMea

Final Measurement:
Detectors: X QP / + AV
Meas Time: 1sec
Subranges: 50
Acc Margin: 6 dB

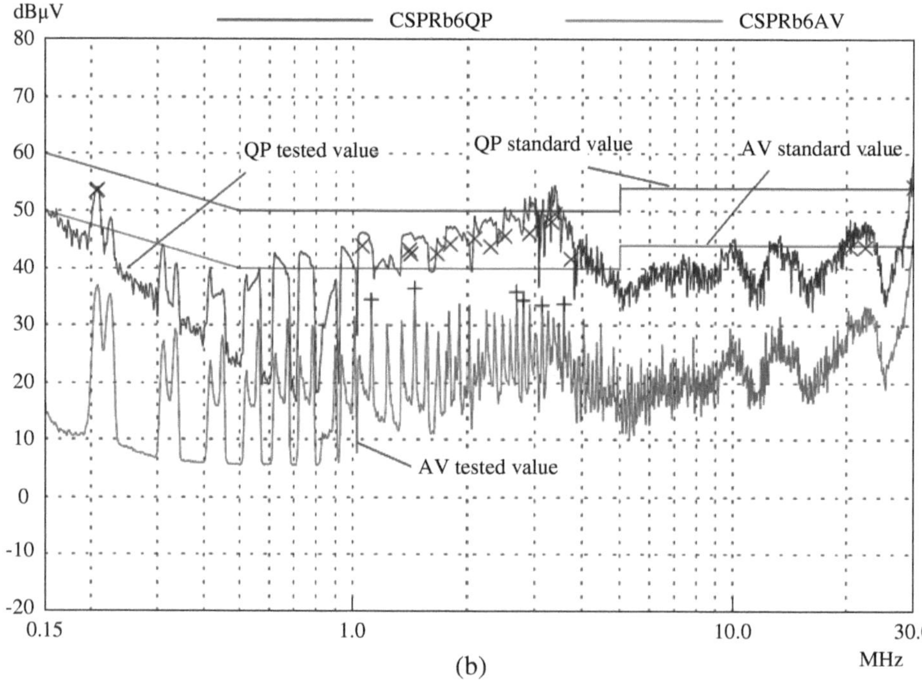

(b)

Figure 10.33 *(continued)*

CONDUCTED EMI

CISPR PUB.22 CLASS B LESS 6dB

31 Oct 2005 10:57

EUT:
Manuf: ASTEC--CDE
Op Cond: 230VAC
Operator:
Test Spec:
Comment: L1 and N tested

Scan Settings (2 Ranges)

Frequencies			Receiver Settings					
Start	Stop	Step	IF BW	Detector	M-Time	Atten	Preamp	OpRge
150kHz	3MHz	0.8%	10kHz	PK+AV	50msec	Auto	OFF	60dB
3MHz	30MHz	0.4%	10kHz	PK+AV	5msec	Auto	OFF	60dB

Transducer	No.	Start	Stop	Name
	11	9kHz	30MHz	LocalMea

Final Measurement: Detectors: X QP / + AV
Meas Time: 1sec
Subranges: 50
Acc Margin: 6 dB

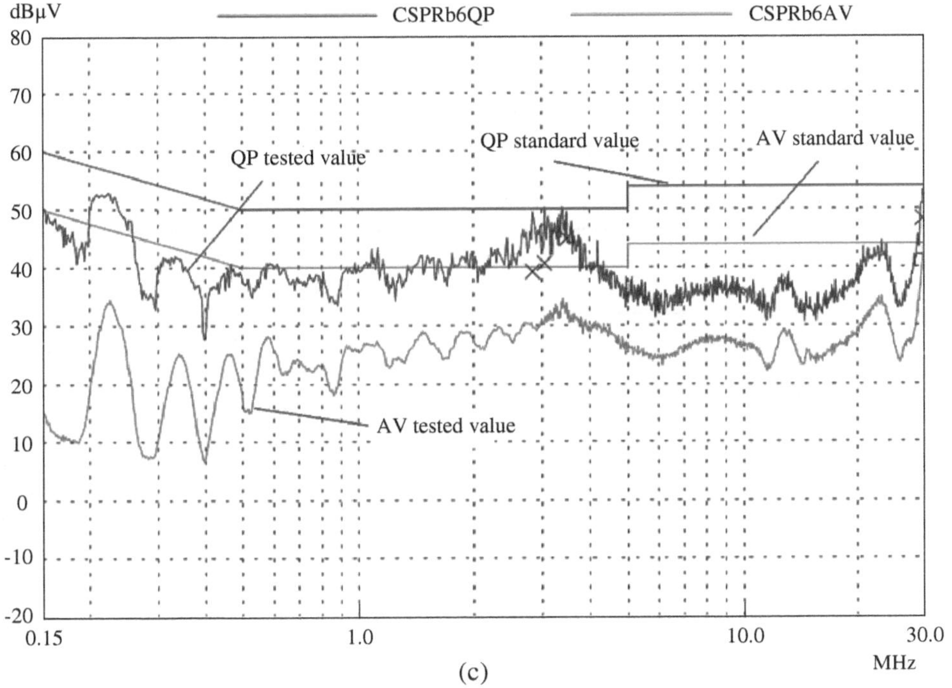

(c)

Figure 10.33 (*continued*)

frequency band, the spectrum peak is slightly lower. In the middle band, the spectrum is more continuous and the peak value is lower. Both QP and AV test values are below the limitation lines, and the EMI test is qualified.

The efficiency and stability tests are performed on the prototype under different loads and two types of input voltage: 115 and 230 V. The test results are listed in Tables 10.4–10.6, and the efficiency curve is in Figure 10.34. Hence we are able to make the following observations from these figures.

Table 10.4 Efficiency and stability test under constant frequency

115 Vac					25% load	50% load	75% load	100% load
	Vo (V) E-load				20.276	20.194	20.122	20.030
	Io (A)				0.823	1.645	2.447	3.258
	Pin (W)				18.651	36.502	54.767	73.421
	Efficiency (%)				89.475	91.006	89.905	88.882
230 Vac		**0 W**	**0.5 W**	**3 W**	**25% load**	**50% load**	**75% load**	**100% load**
	Vo (V) E-load	20.352	20.399	20.331	20.312	20.243	20.083	20.000
	Io (A)	0.000	0.0268	0.153	0.822	1.632	2.453	3.255
	Pin (W)	0.371	0.9513	3.775	18.730	36.691	54.658	72.280
	Efficiency (%)	0.000	57.468	82.401	89.143	90.034	90.139	90.066

Table 10.5 Efficiency and stability test under sinusoidal frequency-spreading

115 Vac					25% load	50% load	75% load	100% load
	Vo (V) E-load				20.263	20.179	20.096	20.020
	Io (A)				0.806	1.627	2.433	3.248
	Pin (W)				18.501	36.408	54.598	73.249
	Efficiency (%)				88.276	90.176	89.552	88.772
230 Vac		**0 W**	**0.5 W**	**3 W**	**25% load**	**50% load**	**75% load**	**100% load**
	Vo (V) E-load	20.348	20.395	20.328	20.255	20.169	20.084	20.000
	Io (A)	0.000	0.0249	0.148	0.816	1.626	2.432	3.248
	Pin (W)	0.371	0.9510	3.785	18.720	36.680	54.440	72.280
	Efficiency (%)	0.000	53.400	79.379	88.280	89.408	89.721	89.873

Table 10.6 Efficiency and stability test under chaotic frequency-spreading

115 Vac					25% load	50% load	75% load	100% load
	Vo (V) E-load				20.258	20.172	20.086	20.012
	Io (A)				0.781	1.586	2.412	3.243
	Pin (W)				18.218	36.202	54.509	73.235
	Efficiency (%)				86.845	88.373	88.879	88.617
230 Vac		**0 W**	**0.5 W**	**3 W**	**25% load**	**50% load**	**75% load**	**100% load**
	Vo (V) E-load	20.348	20.378	20.218	20.213	20.158	20.079	20.000
	Io (A)	0.000	0.0237	0.144	0.784	1.592	2.412	3.223
	Pin (W)	0.371	0.9473	3.778	18.320	36.34	54.420	72.130
	Efficiency (%)	0.000	50.983	77.061	86.501	88.309	88.994	89.367

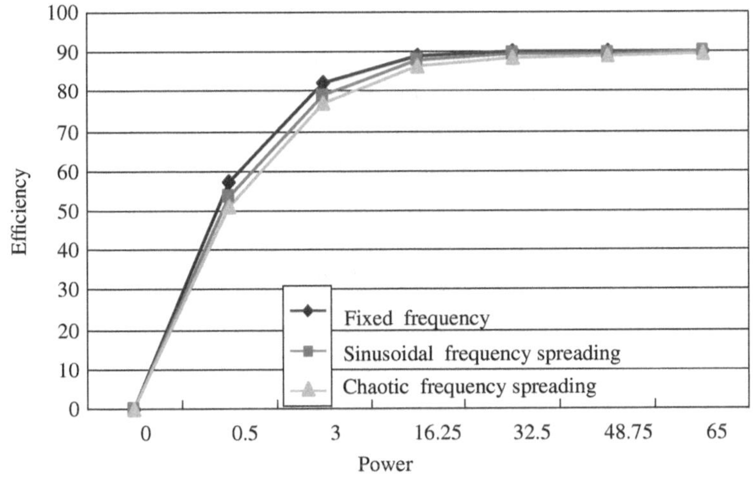

Figure 10.34 Efficiency curve

1. The efficiencies of sine modulation and chaos modulation are slightly lower. The efficiency of chaos modulation is only about 0.6 and 1.7% reduction at full load and middle load respectively.
2. After frequency spreading, the efficiency shows an obvious reduction for light loads.
3. The efficiency of chaos modulation is slightly lower than sine modulation.

We studied these results and found the main factors are as follows.

1. The generating circuits of the sine signal and chaos signal all absorb power from the auxiliary winding of the high frequency transform of the power supply. Therefore, these circuits consume power and cause the efficiency decrease as it should be.
2. The consumed power of the modulation signal generating circuit is usually constant and does not vary with load. Thus, the efficiency varies obviously in light loads compared with full and middle loads. Conversely, the efficiency varies little in full and middle loads.
3. The component quantity of the chaos signal generating circuit is larger than that of the sine signal generating circuit, and the consumed power is also slightly bigger. So, the efficiency of chaos modulation is slightly lower than that of sine modulation.

The efficiency and stability tests under different level loads and two level input voltages are performed on an experimental prototype. The test results show that the prototype works stably in kinds of test conditions. Although the modulation signal generating circuits consume a certain amount of power, the efficiency changes little. So the chaos modulating PWM is feasible in power supplies.

The engineering realization problems of chaotic PWM suppressing the EMI of power electronic converters were studied. Chaotic PWM regulation parameters and chaotic signal choice principles were proposed, and an easy chaotic signal generation

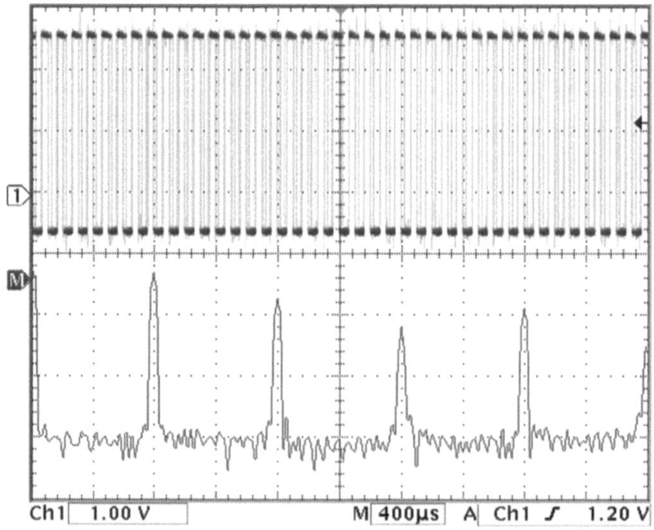

(a)

Figure 10.35 Driving signal (top) and spectrum (down) modulated by chaotic maps with (a) fixed-frequency drive, (b) Rényi map modulated, (c) tent map modulated, (d) Bernoulli map modulated, and (e) Bended up-down map modulated

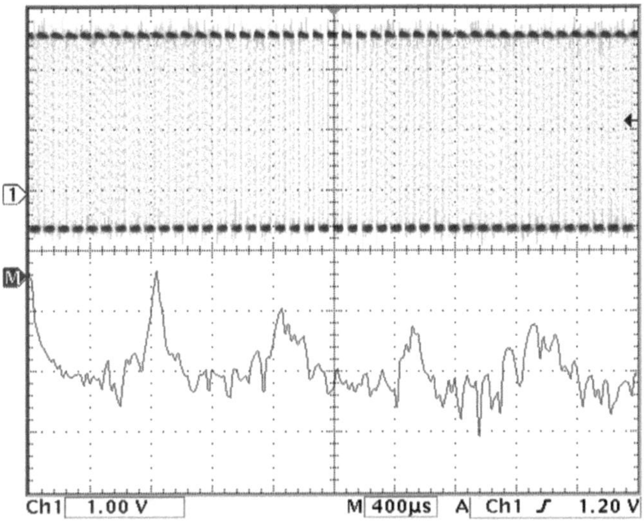

(b)

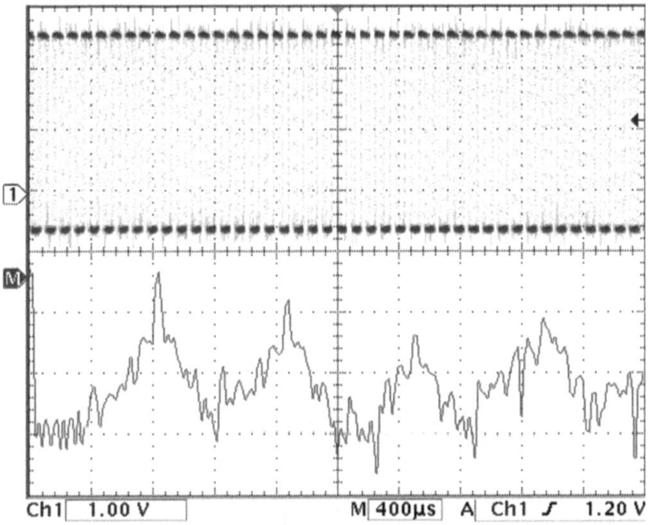

(c)

Figure 10.35 (*continued*)

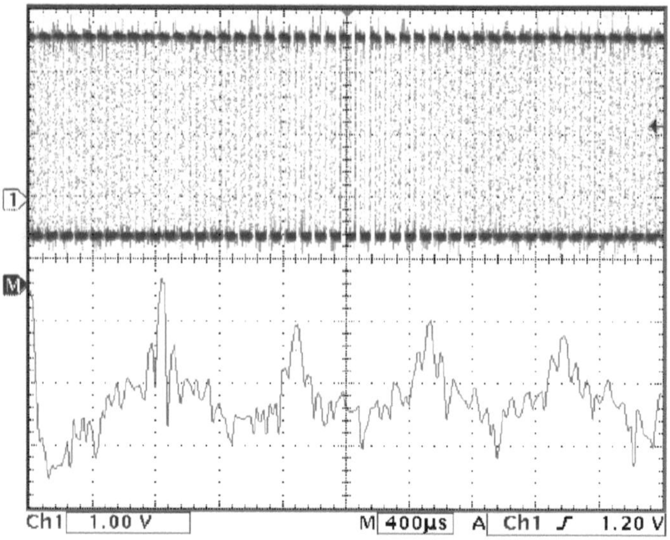

(d)

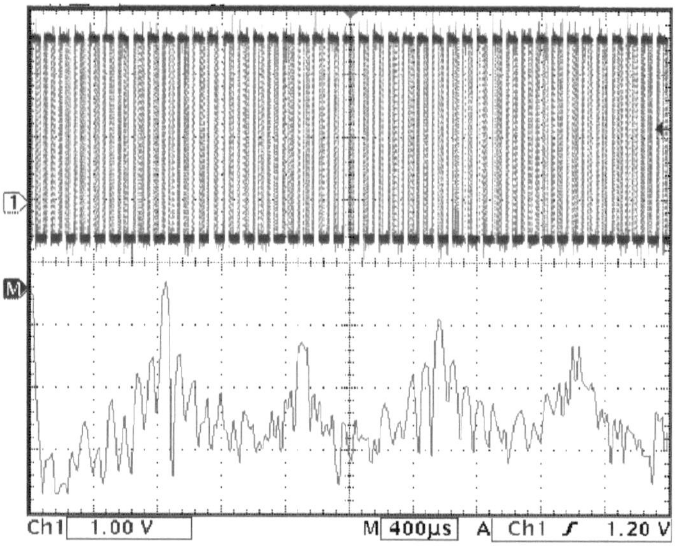

(e)

Figure 10.35 (*continued*)

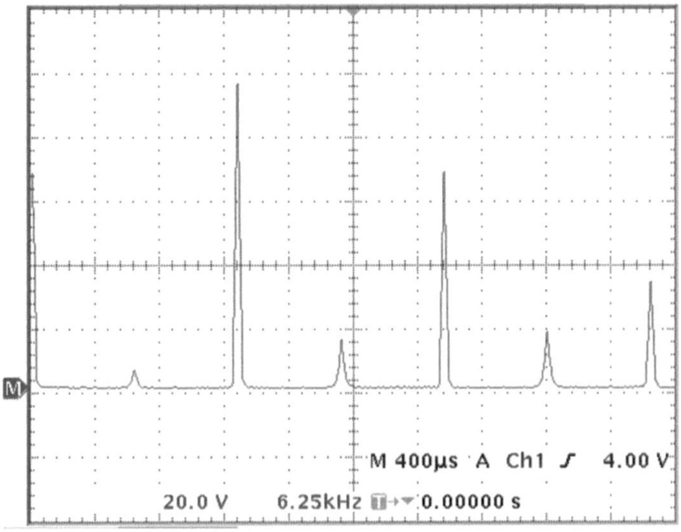

(a)

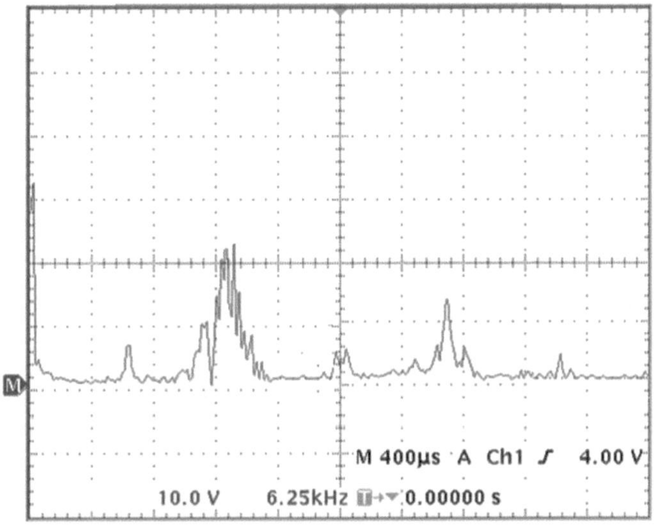

(b)

Figure 10.36 Spectrum of inverter output voltage with (a) fixed-frequency drive, (b) Rényi map modulated, (c) tent map modulated, (d) Bernoulli map modulated, and (e) bended up-down map modulated

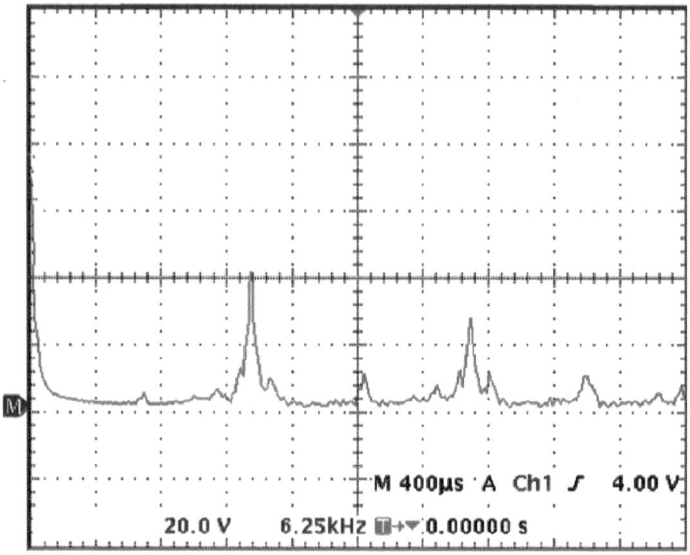

(c)

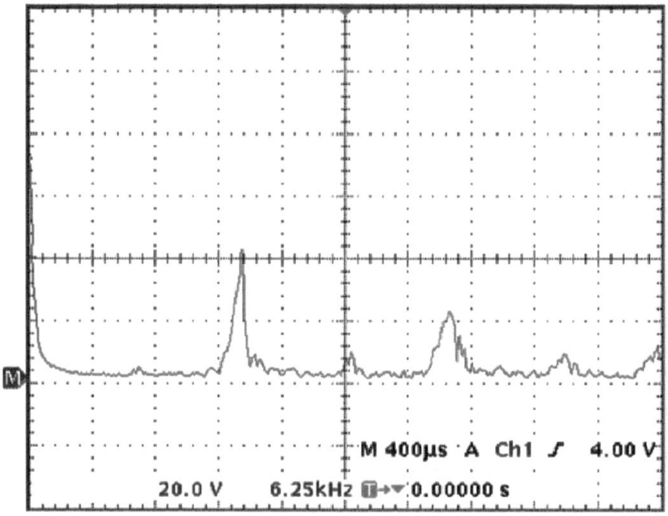

(d)

Figure 10.36 (*continued*)

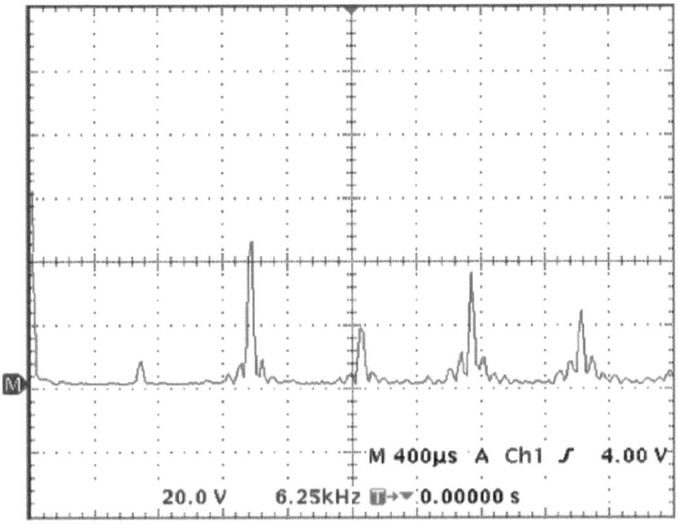

(e)

Figure 10.36 *(continued)*

circuit – a piecewise-linear capacitive chaotic generation circuit – was designed, and the chaotic modulation methods based on the special PWM chip UC3842 and SG3525 were provided. Two experimental prototypes of chaotic modulation boost converter and full bridge inverter were manufactured on this basis. The experiment proved that the proposed chaotic PWM suppressing EMI of power electronic converters is a simple and effective method.

10.6 Characteristics of Chaotic Modulated by Different Chaotic Maps

A chaotic EMI suppressing experiment was also completed on a Space Vector Pulse Width Modulation (SVPWM) inverter, where an Integrated 500 W Power Modular (IPM) IRAMS06UP60A was employed and the main algorithm controller was Advanced RISC Machines (ARM) LPC2132 of Philip@. The carrier frequency was 10 kHz, modulation index was 0.5, modulated wave frequency was 50 Hz, and the frequency changes in the range of 0–20%. The driver waveforms and FFT spectrum of the inverter are shown in Figure 10.35.

Compared with the fixed frequency driver (Figure 10.35a), the spectrum peak after modulation was lower by about 500 mV. The Bended up-down map is more decentralized than other chaotic maps, so we may imagine that the EMI suppressing effects should be worse than others.

The output voltage waveforms and spectrum are shown in Figure 10.36. The spectrum of the fixed drive frequency is concentrated in the fundamental and integral

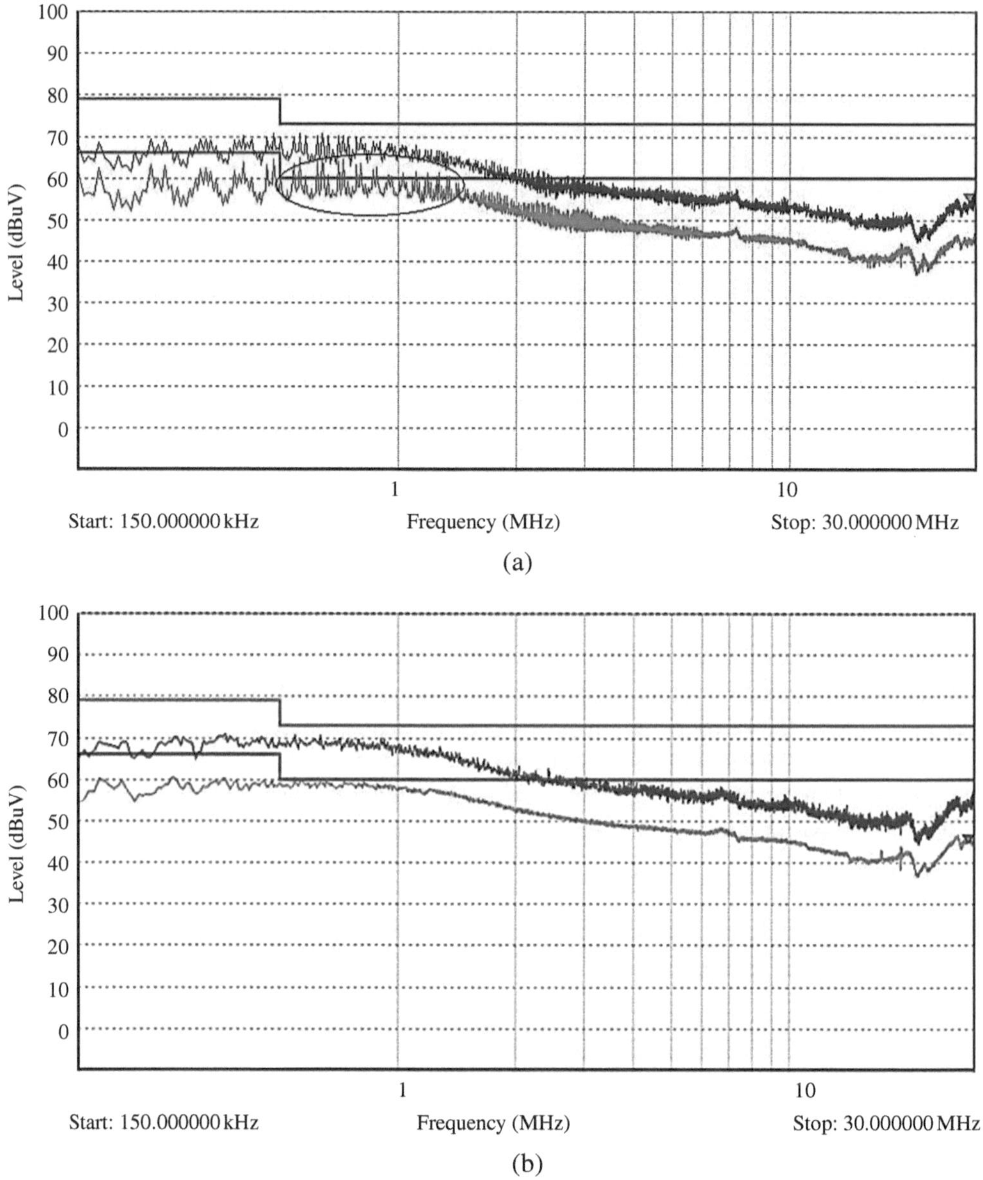

Figure 10.37 Conducted EMI test results of an inverter with (a) fixed-frequency drive, (b) Rényi map modulated, (c) tent map modulated, (d) Bernoulli map modulated, and (e) bended up-down map modulated [18]

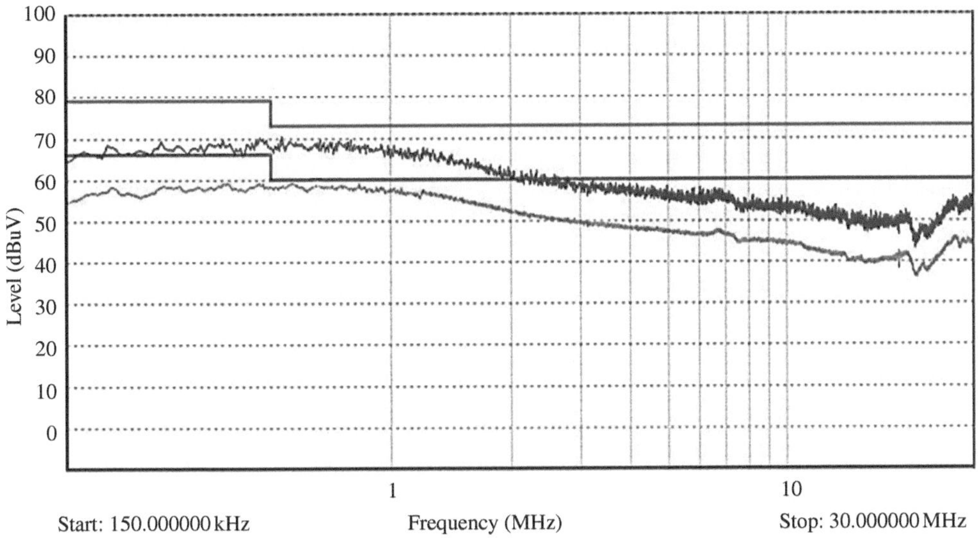

(c)

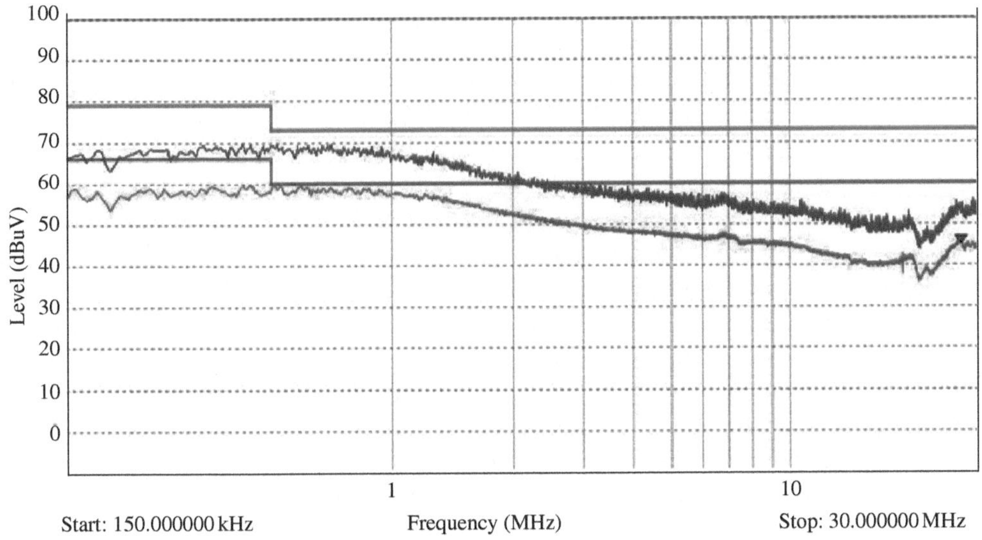

(d)

Figure 10.37 (*continued*)

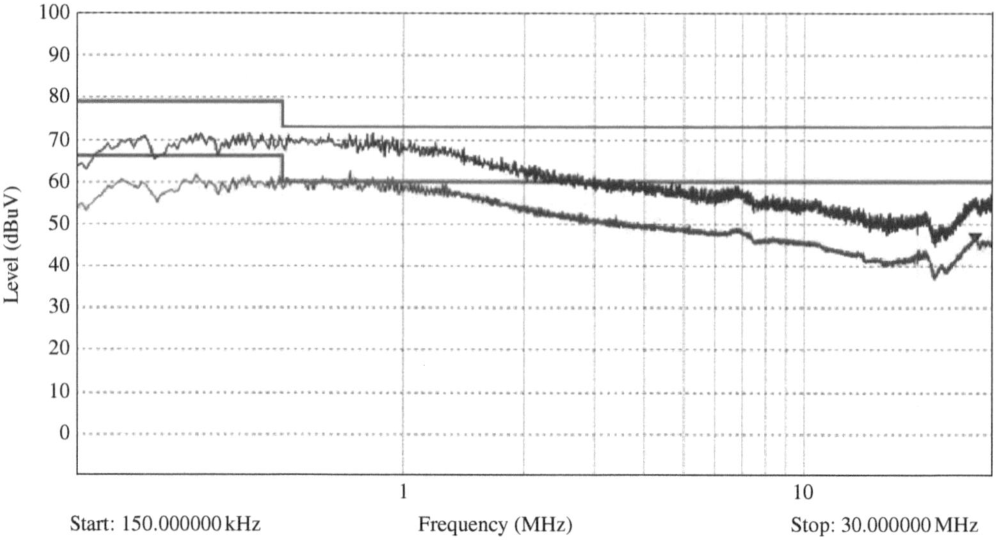

(e)

Figure 10.37 *(continued)*

harmonics. The spectral peaks by chaos modulation are reduced, because the duty cycle keeps constant and the fundamental amplitude does not change.

An EMI conducted interference test was performed in a shielded room by an EMI tester of Rohde & Schwarz, where the CISPER test standard was adapted. The test results of the EMI conducted interference show that in a constant carrier, the AV of EMI interference exceeds the test standard in the frequency band 400 kHz to 1.5 MHz (demonstrated by cycles in Figure 10.37a). After chaotic modulation, the inverter's EMI was improved significantly (demonstrated in Figure 10.37b–e), where the peaks of the test results of AV are lower than the AV standard by Rényi, tent, and Bernoulli piecewise-linear maps modulation respectively. But the EMI suppressing effect by Bended up-down chaotic modulation is worse than other chaotic map modulations. Compared with the driver and output voltage of Figures 10.35 and 10.37, the conclusion will be drawn that the chaotic spectrums reflect the spectral characteristics and EMI of the whole system (Figure 10.37).

10.7 Conclusions

The principle of chaotic PWM suppressing EMI was introduced in this chapter. We provided two key techniques of chaotic PWM for power electronic converters to suppress EMI, including the parameter selection method and chaotic PWM modulation signal method. Then we showed some experimental examples to verify the effect of suppressing EMI by the presented methods. Finally, we analyzed EMI suppression in a commercial switching power supply and the characteristics of chaotic modulated by different chaotic maps.

References

[1] Kuisma, M. (2003) Variable frequency switching in power supply EMI-control: an overview. *IEEE Aerosp. Electron. Syst. Mag.*, **18**(12), 18–22.

[2] Fan, C.X. (1995) *Communication Theory*, National Defense Industry Press, Beijing (in Chinese).

[3] Chen, B.Y. (2006) *RF Communication Circuit*, Science Press, Beijing (in Chinese).

[4] Balcells, J., Santolaria, A., Orlandi, A. *et al.* (2005) EMI reduction in switched power converters using frequency modulation techniques. *IEEE Trans. Electromagn. Compat.*, **47**(3), 569–576.

[5] Lin, F. and Chen, D.Y. (1994) Reduction of power supply EMI emission by switching frequency modulation. *IEEE Trans. Power Electron.*, **9**(1), 132–137.

[6] Tse, K.K., Chung, H.S.-H., Huo, S.Y. and So, H.C. (2000) Analysis and spectral characteristics of a spread-spectrum technique for conducted EMI suppression. *IEEE Trans. Power Electron.*, **15**(2), 399–410.

[7] Tse, K.K., Chung, H.S.-H., Huo, S.Y. and So, H.C. (2002) A comparative study of carrier- frequency modulation techniques for conducted E MI suppression in PWM converters. *IEEE Trans. Ind. Electron.*, **49**(3), 618–627.

[8] Tse, K.K., Ng, R.W.-M., Chung, H.S.H. *et al.* (2003) An evaluation of the spectral characteristics of switching converters with chaotic carrier-frequency modulation. *IEEE Trans. Ind. Electron.*, **50**(1), 171–182.

[9] Setti, G., Mazzini, G., Rovatti, R. and Callegari, S. (2002) Statistical modeling of discrete-time chaotic processes-basic finite-dimensional tools and applications. *Proc. IEEE*, **90**(5), 662–690.

[10] Yang, R. (2007). Chaotic characteristics and EMI suppression of power electronic converters. Doctoral dissertation (in Chinese). South China University of Technology, Guangzhou, China.

[11] Deane, J.H.B., Ashwin, P., Hamill, D.C. and Jefferies, D.J. (1999) Calculation of the periodic spectral components in a chaotic DC-DC converter. *IEEE Trans. Circuits Syst. I*, **46**(11), 1313–1319.

[12] Chua, L.O. (1978) *Introduction to Nonlinear Network Theory*, Robert E. Krieger Press, New York.

[13] Gao, J.F., Ma, X.K. and Luo, X.J. (2001) Hopf bifurcation of nonlinear capacitance double scrolls chaotic circuit. *Trans. Chin. Electrotech. Soc.*, **16**(3), 75–79 (in Chinese).

[14] Orabi, M. and Ninomiya, T. (2003) Nonlinear dynamics of power-factor-correction converter. *IEEE Trans. Ind. Electron.*, **50**(6), 1116–1125.

[15] Gao, J.F., Li, G.M. and Wang, J.K. (2006) From period-doubling to chaos in a new double scroll circuit containing a nonlinear capacitor. *J. Circuits Syst.*, **5**(4), 66–69 (in Chinese).

[16] Wang, X.M. (2009) *Study and Application of Bifurcation and Chaos behavior of Power Electronic Converters Based on Discrete System Dynamics*, South China University of Technology, Guangzhou (in Chinese).

[17] Yang, R., Zhang, B. and Qiu, D.Y. (2009) Time–frequency and wavelet transforms of EMI dynamic spectrum in chaotic converter. *IEEE Trans. Power Electron.*, **24**(4), 1083–1092.

[18] Yang, R. (2007) *Chaotification and EMI Suppression of Power Converter*, South China University of Technology, Guangzhou (in Chinese).

[19] Wang, X.M. and Zhang, B. (2007) Proportional and differential control of a nonlinear capacitor chaos circuit. International Conference on Automation and Logistics, Jinan, China, pp. 2462–2466.

Index